CONTENTS

TABLES

FIGURES

SUMMARY

Carbon dioxide and other atmospheric gases--the so-called "greenhouse gases"--trap some of the sun's heat near the earth, creating the climatic conditions that make life possible. Some investigators are concerned that rising concentrations of these gases, largely resulting from human activities, may cause an increase in the Earth's average temperatures that could have severe economic and ecological effects. Although there is great uncertainty about the extent to which such global warming is likely to occur, what its effects might be, and the costs of efforts to slow the progress of warming, the potential consequences have led to calls for immediate action.

One way to delay or avert global warming would be to reduce the rate at which greenhouse gases are released into the atmosphere. Carbon dioxide contributes about half of the problem caused by greenhouse gases. The contribution of carbon dioxide could become even larger over the next decade as international agreements lead to reduced releases of chlorofluorocarbons, the most reactive gases.

Reducing emissions of carbon dioxide will very likely be a part of any response by governments to the threat of global warming. To the extent that such emissions stem from human activities, their major source is easily identified as the combustion of fossil fuels: coal, oil, and natural gas. Furthermore, emissions of carbon dioxide are relatively easily subjected to government influence or controls, though these may be costly.

The main value of U.S. action to curb emissions of carbon dioxide may be in furthering international efforts. The United States accounts for about 25 percent of world emissions of carbon dioxide--a substantial amount for a single country but still not enough to allow unilateral U.S. action to reduce global warming significantly. Concerted action by many governments would be required if substantial progress is to be made. Several industrialized countries other than the United States have already adopted goals for cutting emissions. Action by the United States could advance these international efforts.

Emissions of carbon dioxide could be reduced through regulatory controls or by using economic incentives such as taxes, subsidies, or marketable permits for the production or use of fossil fuels. Economic incentives generally have an advantage over regulatory controls in that less information would be needed to achieve a given reduction in emissions of carbon dioxide at the lowest cost to society. Well-designed economic incentives would work through the price system to induce changes in consumption and use that would cause carbon dioxide emissions to fall.

Carbon charges--taxes on fossil fuels set according to their carbon contents--are one type of economic incentive that could effectively reduce emissions of carbon dioxide. Carbon charges would cause the prices faced by users of fossil fuels to rise. Since coal has the largest carbon content per unit of energy, its price would rise the most, followed by those of oil and natural gas. The price increases would cause the economy to reduce its overall consumption of energy and to substitute other energy forms that create less carbon dioxide. The changes would be brought about in five ways: by reducing the consumption of all goods, by inducing changes in consumption patterns, by inducing changes in how goods are produced, by encouraging fuel substitution, and by stimulating technological advances. Carbon charges would provide an incentive to adopt these measures in combinations that would reduce carbon emissions from fossil energy combustion in a cost-effective manner.

This study assumes that carbon charges would be phased in over a 10-year period, beginning at $10 per ton of carbon in 1991 and rising to $100 per ton of carbon in 2000 (in 1988 dollars). When fully phased in, the carbon charges would amount to $60.50 per ton of coal, $12.99 per barrel of oil, and $1.63 per thousand cubic feet of natural gas (mcf). These final tax rates would be about half of the projected prices of oil and gas (measured at the wellhead or dock, if imported), but about 2.5 times the projected price of coal (measured at the mine mouth). Because coal has a greater carbon content per unit of energy than the others and because the price of coal per unit of energy is relatively low, a carbon charge would thus affect coal more than the other fuels.

Choosing to impose a carbon charge and deciding on its level would require considering the magnitude of the benefits to be expected, including the direct effects on global warming and the indirect effects such as improved air quality. The direct benefits of carbon charges-- slowing the growth of concentrations of greenhouse gases and delaying global warming--would probably not be felt until well into the twenty-first century. This study does not undertake to estimate the magnitude of the benefits.

EFFECTS OF CARBON CHARGES IN THE FIRST DECADE

Carbon charges would raise the prices of fossil fuels and the prices of products produced using fossil fuels. These price rises would cause producers to try to use less fossil fuel or to shift among fuels so as to reduce costs. Consumers would tend to shift away from products that had risen in price because relatively large amounts of energy were used in their production. This combination of effects would cause the total amount of fossil energy used to decline relative to the baseline levels-- that is, the amount that would have been used in the absence of the carbon charge. Coal use, in particular, would drop because the tax would raise the price of coal more than that of the other fuels and because opportunities exist within the electric utility sector to substitute for coal. Thus, emissions of carbon dioxide would fall as use of fossil fuels declined.

This study uses economic models of energy markets and the U.S. economy to analyze the effects of the carbon charge. The results of two models capturing the shorter-run effects of the tax indicate that emissions of carbon dioxide would fall below baseline levels by an estimated 8 percent to 16 percent by the year 2000. One goal that has been prominent in discussions of policies to combat global warming is stabilizing carbon dioxide emissions at 1988 levels by the year 2000, with further reductions in subsequent years. A tax at the rate specified would result in emissions that might range from 5 percent above to 6 percent below 1988 levels in the year 2000. A higher tax could cause greater reductions in emissions of carbon dioxide. The higher the tax, however, and the more quickly it was imposed, the greater would be the resulting economic dislocation and the possible slowing of economic growth.

Effects on Federal Revenues

Additional revenues generated by the carbon charge examined in this study could be about $12 billion in calendar 1991, when the phase-in period of the tax is assumed to begin (revenue estimates represent the net effect on the federal deficit). In 2000, when the tax rate would reach $100 per ton of carbon, revenues are estimated to be between $110 billion and $120 billion (all figures are in 1988 dollars).

Effects on Energy Use

Demand for energy in the residential, commercial, industrial, and transportation sectors (that is, the end-use sectors) would fall relative to the baseline when carbon charges were applied. Shifting away from coal would be most pronounced among electric utilities, where about 80 percent of the current production of coal is used. Carbon charges could make some coal plants more expensive to operate in the same power pool. Utilities that have both gas- and coal-fired generating capacity would tend to increase their use of the gas-fired plants and cut back on their use of the coal plants in the near term. Over a 10-year period, electric utilities would take account of these increased generation costs when planning for additional capacity.

Industrial Effects

The coal industry would be relatively hard hit by carbon charges. One model used in this study indicated that the industry could suffer about a 25 percent drop in output by the end of the century compared with baseline levels. Also, generally speaking, industries that are relatively intensive in their use of energy would be affected more adversely than other industries. For example, industries producing steel, clay, glass, rubber, plastic, and other chemical products are relatively energy-intensive and could face reductions in size if carbon charges were imposed. In contrast, producers of foods, fibers, and other agricultural products might expand their sales because the relative prices of their products would tend to fall.

Aggregate Economic Effects

Carbon charges of the size assumed for this study would have discernible effects on national income, the general level of prices, and the balance of trade. One estimate described in the study is that the gross national product (GNP) would be about 2 percent less by 2000 than without the charge (2 percent of gross national product is now roughly $100 billion). If the charge was set at $100 per ton initially and not phased in, the immediate effect on the economy would be far more severe, possibly leading to stagnation and significant unemployment for several years.

A slowdown in economic performance would not be unique to carbon charges, but the effect of carbon charges might be more difficult for the economy to digest than other changes in fiscal policy. Most changes in revenue or spending policy intended to reduce the federal budget deficit could also contribute to a slowing of the economy. Such an effect could be lessened if the Federal Reserve eased interest rates and stimulated the economy. But carbon charges--unlike deficit reduction measures that do not directly increase prices--would confront the Federal Reserve with a difficult choice: whether to fight the inflationary effects of the charges or to mitigate their effects on economic growth. In the past, the Federal Reserve has chosen to allow some increase in unemployment in order to reduce the inflationary impacts of increases in energy prices, and it might choose to do so again. This short-run risk could be avoided by phasing the charges in and using some potential revenues to reduce other taxes or increase expenditures in selected areas.

Effects on International Trade

In examining the short- and intermediate-term effects of carbon charges, the study assumes that they would be imposed only by the

United States and not by other countries. If so, the charges would affect the competitiveness of some U.S. industries in international trade--particularly those using relatively large amounts of energy in their production processes. The charges might affect the composition of U.S. trade without significantly affecting the overall trade deficit. The trade deficit exists largely because of macroeconomic factors such as the level of national saving and the size of the federal deficit. To the extent that carbon charges contributed to reducing the federal deficit, they would help to reduce the trade deficit.

U.S. industries whose products embody--directly and indirectly--a large content of carbon-bearing fuels would lose out, while other less carbon-intensive U.S. industries competing in import or export markets could gain from the carbon charges. However, these effects on the composition of trade would be lessened if major trading partners were to adopt carbon charges or similar measures.

EFFECTS OF CARBON CHARGES OVER A LONGER PERIOD

As compared with a short period of years, considerably larger changes in energy consumption would be achieved after sufficient time had passed to allow changes in the stock of buildings and durable equipment, movement of workers and other resources from declining to growing industries, the introduction of new carbon-saving technologies, and changes in the purchasing patterns of consumers. One set of estimates done for this study suggests that the results of these longer-term changes would be to reduce carbon dioxide emissions by 36 percent below the baseline, at a cost of less than 1 percent of GNP. But this accomplishment in terms of carbon dioxide emissions would require massive changes in energy markets.

Reductions would be required in every form of energy consumption in every end-use sector, with the largest percentage decreases occurring in coal consumption (perhaps reducing coal production to half its current level). As end-use consumption of electricity declined relative to the baseline, demand for fossil fuel for generating electric power would also fall, again with the bulk of the reduction concentrated in coal, but with no other form of fossil energy benefiting from an ability to substitute for coal.

Over the longer term, consumers could give up 20 percent to 30 percent of their direct energy consumption if faced with prices that were higher by 30 percent or so for oil, gas, and electricity. But they would also change their demands for all other goods. This broad change in consumer purchasing habits would allow the substantial reductions in commercial and industrial energy use that would contribute to reduced emissions. Since makers of goods tend to substitute components made with less direct or indirect energy content for those made with more direct energy content, the result would be a pattern of dramatic increases and decreases in output among industrial sectors.

Even the year 2000 might be too soon for all the implicit adjustments to work themselves out. Whatever the ultimate ability of the economy to rearrange itself, these adjustments would probably be accompanied by a difficult and painful transition period. Moreover, given the likely timing of increases in demand for energy, the exhaustion of gas resources, and the availability of new carbon-free energy sources, maintaining stable levels of carbon emissions beyond the year 2000 might also be very difficult.

The interplay among these three factors--changes in energy demand, in the availability of natural gas, and in the rate of introducing new energy technologies--would be the key to the outcome in carbon dioxide emissions over the next century. While developments cannot be foreseen with much certainty, two scenarios may bound the likely possibilities. In the first scenario, assuming that more and more ways to improve the efficiency of energy use are found, and that new energy technologies become available rapidly and are widely adopted, carbon dioxide emissions from the United States might roughly double, even without any purposeful policy, by the year 2100. In this case, the goal of holding carbon dioxide emissions 20 percent below 1988 levels in the year 2100 could be achieved with a carbon charge of $100 per ton and at an annual cost of about 1 percent of GNP during this period.

The second scenario assumes that energy efficiency improves more slowly, at a rate more consistent with historical experience, and that new energy technologies appear at times consistent with current projections of the electric utility industry. In this case, carbon dioxide emissions from the United States could reach levels three times their current levels by the year 2100. Charges might have to reach $250 or more per ton by 2100 to achieve the goal of a 20 percent reduction. A serious problem could be the unavailability of new energy sources on a sufficiently wide scale in the years around 2020, requiring carbon charges as high as $400 per ton to hold emissions down. The annual cost of such measures is estimated to average 3 percent of GNP through the next century.

CONCLUSIONS

Carbon charges can be effective in reducing carbon dioxide emissions from fossil fuel combustion in the United States. Rapidly imposing large charges could be hazardous for the economy, but the costs could be held to a loss of 1 percent to 2 percent of GNP annually during the first decade by phasing in the charges and taking offsetting actions to mitigate the contractionary effects of the charges. Over the longer term, carbon charges of $100 per ton could hold the level of GNP at least 1 percent lower than without the charges. To prevent growth in carbon dioxide emissions after 2000 would require even higher charges.

Moreover, achieving substantial reductions in global carbon dioxide emissions would require multilateral action. Prospects for such action seem good, however, since a number of other industrialized countries have expressed a willingness to curb emissions. If the United States unilaterally imposed a charge of $100 per ton, it might delay the doubling of global carbon dioxide concentrations by only a few years, while a charge of that amount imposed by all countries might buy almost two decades. (A doubling of the atmospheric concentration of carbon dioxide from preindustrial levels has become a convenient point of reference for analyzing the extent and timing of global warming.) Multilateral adoption of charges rising to $300 by 2100 might delay the doubling of concentrations into the 22nd century.

It is unlikely that the costs of reducing carbon dioxide emissions could be lessened by adopting regulatory measures rather than economic incentives. Carbon charges would provide balanced incentives for energy consumers to pursue all the different ways in which they could reduce their reliance on carboniferous fuels. Only economic incentives can bring forth so wide a range of possible responses. The incentives might need to be supplemented, however, by measures designed to address imperfections in the market that could hinder or prevent some of these responses.

The ultimate question is whether the costs imposed by carbon charges, or any other program equally effective in controlling carbon dioxide emissions, would be worth bearing in order to delay the increase in carbon dioxide concentrations. Uncertainties surrounding both the benefits and the costs of such policies make answering this question difficult. If governments wait until all uncertainties are eliminated, they may sacrifice an important opportunity to deal with the problem. Preventive measures--including reductions in carbon dioxide emissions--are likely to be more effective if taken in the near term. However, the costs of reducing the consumption of fossil energy suggest that there may be merit in looking into other ways of dealing with global warming.

Some other greenhouse gases are more reactive than carbon dioxide--that is, more effective in causing global warming--and may potentially be cheaper to control. Some analysts have suggested that complete elimination of chlorofluorocarbons should be considered before drastic reductions in fossil energy use are undertaken. Another atmospheric trace gas whose control would be worth examining is methane. Cultivating plants and forests that help to remove carbon dioxide from the atmosphere would be another approach to the problem. Ways may be found to improve energy efficiency, and to develop new technologies for using forms of energy that do not release carbon dioxide emissions. Finally, governments could assist the economy in adapting to long-run changes in climate--changes that may only be delayed by even the most effective carbon dioxide reduction program.

33-690 O - 90 - 2

CHAPTER I

GLOBAL WARMING--THE PROBLEM AND

A STRATEGY FOR ABATEMENT

A tax on oil, gas, and coal could provide an incentive to reduce the use of these fossil fuels and, consequently, slow down the emissions of carbon dioxide into the atmosphere. Carbon dioxide is one of a number of atmospheric gases--including water vapor, ozone, methane, nitrous oxide, and chlorofluorocarbons (CFCs)--that trap the sun's heat energy near the earth. Without this "greenhouse effect" the earth would not support life.

There is wide agreement that unless significant actions are taken, the concentrations of these heat-trapping gases in the atmosphere will continue to increase as the world population grows and the world economy expands. There is less agreement within the scientific community as to how much the increasing concentrations of greenhouse gases will cause global temperatures to rise, and how rapidly. Further, immense uncertainty exists about the effects that such increases in temperature would have on the climate and on human life.

The economic, environmental, and social effects of relatively rapid warming and the resulting changes in climate could be quite serious. Some analysts predict chronic drought, rising sea levels, and destructive storms--climatic effects that could diminish worldwide food supplies, inundate coastal regions, and cause wide-scale extinction of plants and animals. Others, however, argue that natural mechanisms will temper the effects of higher concentrations of greenhouse gases. Increased cloud cover or greater absorption of greenhouse gases by the oceans could mitigate the effects on temperatures and climate. Still other scientists are concerned that feedback effects and irreversibilities could cause warming, once its effects were felt, to become an increasingly severe problem.

Global warming has aroused interest well beyond the scientific community. The Intergovernmental Panel on Climate Change (IPCC) was formed in 1988 to study the scientific information and to assess strategies of response. This panel, formed by the United Nations Environment Program and the World Meteorological Organization, is the principal international body addressing the issue. The full report of the IPCC is expected to be submitted in the fall of 1990, after which the United Nations may seek to encourage international action (see the box on page 4).

Carbon dioxide is but one of several greenhouse gases that could be brought under control in a comprehensive strategy to avert or delay global warming. Moreover, an appropriate mix of policies might include measures to adapt to the effects of global warming in addition to delaying its onset. The range of possible policy responses is cited in this chapter, but the analysis presented later in this study is limited to the effects of charges on emissions of carbon dioxide.

POLICY RESPONSES TO THE THREAT OF GLOBAL WARMING

Governments can respond in three general ways to the perceived threat of global warming, or to the problems caused by global warming. They can take preventive or abatement actions, curative actions, and adaptive actions. An additional response would be to fund research that would help assess the severity of the potential problem as well as assist in developing appropriate ways of responding.

Preventive or abatement actions would slow the accumulation of greenhouse gases in the atmosphere. The policy discussed in this study is an example of such an action. Emissions of greenhouse gases other than carbon dioxide could also be limited. Emissions of chlorofluorocarbons are already being brought under control, primarily because of their effect on stratospheric ozone. Emissions of methane, another significant greenhouse gas, are not yet subject to regulation. Actions to maintain and expand forests, which absorb large amounts of carbon dioxide, would also be included in this category of policy response.

Some preventive actions could have near-term benefits unrelated to their effects on global warming. Less burning of fossil fuels, for example, could improve air quality. Encouraging energy conservation could improve the management of depletable natural resources. All such serendipitous benefits--especially those that would be realized immediately--would enhance the attractiveness of the action. Also, since actions to prevent or delay global warming would have a gradual, cumulative effect, starting them early would increase their effectiveness.

Preventive actions would almost certainly require international cooperation. Actions by a single country, even one as large as the United States, would be of only limited effectiveness.

Curative actions are ways to ward off or reverse the effects of increased atmospheric concentrations of greenhouse gases. Several have been proposed, including reducing the amount of energy retained near the earth by taking measures to reflect solar energy. Road surfaces might be painted white, or roofs might be covered with very reflective material. These measures may sound far-fetched at first, but they could be put into effect over the next 50 years, a period during which global warming could be becoming a problem.

Adaptative actions would reduce the damaging effects of warming rather than affect the rate of warming itself. Adaptation could include moving populations or building dikes in coastal areas, changing water delivery systems, or changing agricultural practices or crops in some regions. If, as many scientists believe, current accumulations of greenhouse gases have already committed the Earth to some warming, adaptation to climate change will have to occur whether or not other actions are taken.

Adaptive actions differ from preventive actions in that they can be delayed until the effects of global warming are actually felt. Also, individuals or local governments could adapt or assist in adapting to climate changes. In many cases, international cooperation might not be necessary.

A SCIENTIFIC ASSESSMENT OF GLOBAL WARMING

A recently published report of a working group of the Intergovernmental Panel on Climate Change summarizes one view of scientists about global warming. The executive summary of this report is reprinted below:[1]

We are certain of the following:

o There is a natural greenhouse effect which already keeps the Earth warmer than it would otherwise be.

o Emissions resulting from human activities are substantially increasing the atmospheric concentrations of the greenhouse gases: carbon dioxide, methane, chlorofluorocarbons (CFCs) and nitrous oxide. These increases will enhance the greenhouse effect, resulting on average in an additional warming of the Earth's surface. The main greenhouse gas, water vapour, will increase in response to global warming and further enhance it.

We calculate with confidence that:

o Some gases are potentially more effective than others at changing climate, and their relative effectiveness can be estimated. Carbon dioxide has been responsible for over half the enhanced greenhouse effect in the past, and is likely to remain so in the future.

o Atmospheric concentrations of the long-lived gases (carbon dioxide, nitrous oxide and the CFCs) adjust only slowly to changes in emissions. Continued emissions of these gases at present rates would commit us to increased concentrations for centuries ahead. The longer emissions continue to increase at present day rates, the greater reductions would have to be for concentrations to stabilize at a given level.

o The long-lived gases would require immediate reductions in emissions from human activities of over 60 percent to stabilize their concentrations at today's levels; methane would require a 15 percent to 20 percent reduction.

Based on current model results, we predict:

o Under the IPCC Business-as-Usual (Scenario A) emissions of greenhouse gases, [there will be] a rate of increase of global mean temperature during the next century of about 0.3°C per decade (with an uncertainty range of 0.2°C to 0.5°C per decade); this is greater than that seen over the past 10,000 years. This will result in a likely increase in global mean temperature of about 1°C above the present value by 2025 and 3°C before the end of the next century. The rise will not be steady because of the influence of other factors.

o Under the other IPCC emission scenarios which assume progressively increasing levels of controls, [there will be] rates of increase in global mean temperature of about 0.2°C per decade (Scenario B), just above 0.1°C per decade (Scenario C) and about 0.1°C per decade (Scenario D).

o That land surfaces warm more rapidly than the ocean, and high northern latitudes warm more than the global mean in winter.

1. Intergovernmental Panel on Climate Change, *Policy Makers Summary of the Scientific Assessment of Climate Change. Report to the IPCC from Working Group I* (June 1990). First-person references are to Working Group I.

o   That regional climate changes [will be] different from the global mean, although our confidence in the prediction of the detail of regional changes is low. For example, temperature increases in Southern Europe and central North America are predicted to be higher than the global mean, accompanied on average by reduced summer precipitation and soil moisture. There are less consistent predictions for the tropics and the southern hemisphere.

o   That under the IPCC Business as Usual emissions scenario, [there will be] an average rate of global mean sea level rise of about 6 cm per decade over the next century (with an uncertainty range of 3 - 10 cm per decade), mainly due to thermal expansion of the oceans and melting of some land ice. The predicted rise is about 20 cm in global mean sea level by 2030, and 65 cm by the end of the next century. There will be significant regional variations.

There are many uncertainties in our predictions particularly with regard to the timing, magnitude and regional patterns of climate change, due to our incomplete understanding of:

o   Sources and sinks of greenhouse gases, which affect predictions of future concentrations;

o   Clouds, which strongly influence the magnitude of climate change;

o   Oceans, which influence the timing and patterns of climate change; and

o   Polar ice sheets, which affect predictions of sea level rise.

These processes are already partially understood, and we are confident that the uncertainties can be reduced by further research. However, the complexity of the system means that we cannot rule out surprises.

Our judgement is that:

o   Global mean air temperature has increased by 0.3°C to 0.6°C over the last 100 years, with the five global-average warmest years being in the 1980s. Over the same period global sea level has increased by 10-20 cm. These increases have not been smooth with time, nor uniform over the globe.

o   The size of this warming is broadly consistent with predictions of climate models, but it is also of the same magnitude as natural climate variability. Thus the observed increase could be largely due to this natural variability; alternatively this variability and other human factors could have offset a still larger human-induced greenhouse warming. The unequivocal detection of the enhanced greenhouse effect from observations is not likely for a decade or more.

o   There is no firm evidence that climate has become more variable over the last few decades. However, with an increase in the mean temperature, episodes of high temperatures will most likely become more frequent in the future, and cold episodes less frequent.

o   Ecosystems affect climate, and will be affected by a changing climate and by increasing carbon dioxide concentrations. Rapid changes in climate will change the composition of ecosystems; some species will benefit while others will be unable to migrate or adapt fast enough and may become extinct. Enhanced levels of carbon dioxide may increase productivity and efficiency of water use of vegetation. The effect of warming on biological processes, although poorly understood, may increase the atmospheric concentrations of natural greenhouse gases.

Federally supported research is another response to global warming. So far, it has been the only response of the U.S. government to this potential problem. Research can be of three types. First, study of the phenomenon itself would clarify what is likely to happen if no direct actions to avert warming are taken. Research would help to reduce differences of opinion as to the wisdom of taking direct and costly actions. Second, scientific research could lead to technological developments that would make it less costly to avert global warming. Policy research would help define the least costly ways of averting global warming. Third, research could find ways to make the effects of global warming less damaging; for example, it could aid in adapting to higher temperatures and climatic changes.

THE CONTRIBUTION OF CARBON DIOXIDE EMISSIONS TO GLOBAL WARMING

Carbon dioxide, primarily from combustion of fossil fuels, contributes importantly to the potential problem of global warming. Carbon dioxide is estimated to constitute about half of the increase in greenhouse gases that create the potential for global warming (see Figure 1).

Figure 1.
The Contribution of Different Greenhouse Gases to the Change in Global Radiative Forcing, 1980-1990

SOURCE: J. Hansen and others, "Regional Greenhouse Climate Effects," in *Coping with Climate Change, Proceedings of the Second North American Conference on Preparing for Climate Change: A Cooperative Approach* (December 1988).

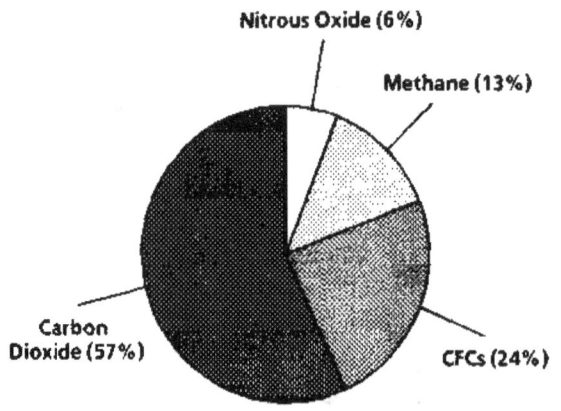

It could account for an even larger share over the next decade as international agreements bring about reductions in the release of CFCs, the most reactive gases.

The United States accounts for about 25 percent of world emissions of carbon dioxide resulting from human activities, which cause emissions of about 20 billion to 28 billion tons of carbon dioxide per year. Although these emissions are modest compared with natural carbon flows, they have apparently tipped the balance between natural sources and sinks (areas such as forests that absorb carbon dioxide), thus increasing the atmospheric concentrations of carbon dioxide. According to the geological record, the concentration of carbon dioxide fluctuated between 180 and 290 parts per million (ppm) over the 160,000 years prior to 1700. In the eighteenth century--at the beginning of the industrial revolution--the level of carbon dioxide was in the range of 250 ppm to 290 ppm. Measures of concentrations of carbon dioxide in polar ice cores indicate concentrations of 300 ppm by 1900. By 1950, they had risen to 315 ppm, and by 1988 to about 350 ppm. While scientists disagree over whether or not this level will affect global climate, the potential effects of ever-rising concentrations of carbon dioxide have prompted wide concern.

Projected Emissions of Carbon Dioxide

Projections of future global emissions of carbon dioxide indicate that atmospheric concentrations are likely to rise rapidly over the next several decades, primarily as a result of increasing use of fossil fuels. Fossil resources now provide almost 90 percent of global annual commercial energy needs.[1] Unless the overall reliance on fossil fuels can be reduced significantly, increases in global energy use will continue to cause carbon dioxide concentrations to rise.

This study uses projections from the U.S. Energy Information Administration's *Annual Energy Outlook 1989* as a baseline for U.S.

---

1.   Fossil energy provides nearly 90 percent of commercial energy in the United States as well. See World Resources Institute and the International Institute for Environment and Development, *World Resources 1988-1989, An Assessment of the Resource Base That Supports the Global Economy* (New York: Basic Books, Inc., 1988), Table 7.1, p 110.

demand for fossil fuels through 2000. The levels of emissions of carbon dioxide consistent with the fuel uses in the baseline were calculated to generate a carbon dioxide baseline, summarized in Table 1. Following this baseline, U.S. emissions of carbon dioxide would rise to a rate 14 percent above 1988 rates by the year 2000.

Choosing an appropriate baseline for the years beyond 2000 is difficult. Such projections depend on assumptions as to rates of economic growth in the United States and the rest of the world, the speed at which reserves of fossil fuels are exhausted, and the rate of introducing new energy sources or new technologies that conserve energy. The results of two analyses examining the effects of carbon charges through 2100 are reported later in this study (see Chapter III). An important difference between the two analyses lies in their baseline projections of emissions of carbon dioxide.

TABLE 1.   PROJECTED USE OF FOSSIL FUELS AND EMISSIONS OF CARBON DIOXIDE IN THE UNITED STATES, 1988-2000

	1988	1989	1990	1991	1992	1993	1994	1995	1996	1997	1998	1999	2000
Use of Fossil Fuels (Quadrillions of Btus) Petroleum products	33.7	33.9	34.4	34.7	35.0	35.1	35.1	35.2	35.7	35.8	35.9	36.4	36.8
Natural gas	18.3	18.5	18.5	18.6	18.6	18.9	19.2	19.5	19.7	20.1	20.5	20.6	20.9
Coal	18.8	19.0	19.1	19.3	19.8	20.2	20.6	21.0	21.0	21.4	21.8	22.2	22.5
Emissions of Carbon Dioxide (Billions of tons)[a]	5.7	5.8	5.8	5.9	6.0	6.0	6.1	6.1	6.2	6.3	6.3	6.5	6.5

SOURCES: Congressional Budget Office projections of use of fossil fuels from Energy Information Administration, *Annual Energy Outlook 1989*. Carbon dioxide coefficients adapted by CBO from Gregg Marland, "Carbon Dioxide Emission Rates for Conventional and Synthetic Fuels," *Energy*, vol. 8, no. 12 (1983), pp. 981-992.

a.   Estimates of carbon dioxide emissions are calculated by applying coefficients to projections of fossil fuel use. The coefficients, expressed as billions of tons of CO_2 per quadrillion Btus, are: coal (.101), oil (.082), and gas (.058).

Targets for Reducing Emissions of Carbon Dioxide

Various goals for reducing worldwide emissions of carbon dioxide have been suggested. An international conference in Toronto, Canada, in June 1988, set an initial goal of reducing world emissions of carbon dioxide by 20 percent below 1988 levels by the year 2005. The Intergovernmental Panel on Climate Change is examining the feasibility of a target of stabilizing emissions by 2000, and then reducing them 20 percent by 2005.

Choosing targets should involve weighing the costs of reducing emissions against the benefits to be achieved. This study estimates some of the costs of achieving significant reductions in emissions of carbon dioxide (bringing emissions nearly back to 1988 levels by the year 2000) by taxing the carbon content of fuels. Little work has been done to quantify the benefits.

CARBON CHARGES AS A MEANS OF REDUCING EMISSIONS OF CARBON DIOXIDE

Carbon charges are taxes on fossil fuels--coal, oil, and natural gas--that are set according to the carbon content, or the potential to release carbon dioxide, of each fossil fuel. Such taxes would reduce the rate of use of fossil fuels, and hence the rate of emissions of carbon dioxide and the accumulation of carbon dioxide in the atmosphere. To be workable, carbon charges would have to be imposed on all forms of fossil energy, whether of domestic origin or imported.

This study looks at some of the economic costs of carbon charges. It examines how emission rates would respond to these taxes in both the near term and the long term, and how the taxes and the resulting reduced rates of use of fossil fuels would affect national income.

Any action by the United States is likely to become part of an international effort to reduce emissions. Several other industrialized countries have already adopted goals for reducing their emissions of carbon dioxide. Indeed, such an international effort would be needed to reduce atmospheric concentrations of carbon dioxide significantly. Even relatively large reductions by any single country are unlikely to make a significant difference in worldwide emissions. For example, bringing U.S. emissions back to 1988 levels in the year 2000--roughly a 12 percent cut from projected levels--would reduce world emissions by less than 3 percent.

This study nevertheless focuses on carbon charges introduced in the United States alone. Since some of the effects of carbon charges on energy markets and the economy would depend on what other nations were doing, it is necessary to make assumptions about whether charges would be imposed unilaterally by the United States or as part of an international agreement. Since no such agreement now exists, most of the analysis in this study assumes that the United States imposes charges unilaterally. The results can be viewed as a rough estimate of the potential contribution by the United States to international efforts to reduce emissions. As Chapter III shows, however, the actual U.S. contribution would depend on the presence or absence of emission-reducing policies abroad.

Fossil fuels are very important to the U.S. economy, and not surprisingly this study shows that it would be quite costly to impose a tax large enough to have a substantial effect in reducing rates of use of fossil fuels and rates of emissions of carbon dioxide. For example, a tax large enough to restrain U.S. emissions of carbon dioxide to 1988 levels in the year 2000 could cost the economy from 1 percent to 2 percent of gross national product by the year 2000 (1 percent of the gross national product being about $50 billion).

Reducing emission levels more drastically--for example, to 20 percent below 1988 levels by 2000 or thereabouts--would require stringent and immediate reductions in carbon dioxide emissions and would entail greater losses to the economy. Carbon charges implemented rapidly enough to cause such large reductions in emissions could result in economic stagnation and significant unemployment for several years. One portion of these economic losses in the near term would be the costs incurred in transition from an economy highly dependent on fossil fuels to one less dependent; another portion would be the shock effect of suddenly increasing energy prices.

Would such a policy be worth the cost? The benefits of averting or delaying global warming are not estimated in this study. It can be inferred from the costs of the policies examined, however, that the consequences of global warming would have to be severe to justify the action.

The costs of a carbon charge policy would begin immediately, but the principal benefits of the policy would not be felt for a long time. Balancing the costs and benefits of a policy that would not be realized for many years and that would affect future generations has been a subject of much debate, and is beyond the scope of this paper. But not all the benefits of a carbon charge policy need wait upon the distant future. Reducing the use of fossil fuels and encouraging the development of other forms of energy could improve air quality immediately.

THE EFFECTS OF CARBON CHARGES

Carbon charges would reduce emissions of carbon dioxide by discouraging the use of fossil fuels or by encouraging users to switch toward a fossil fuel that is lower in carbon content. Since the use of fossil fuels is pervasive in the U.S. economy, the changes in fuel prices caused by carbon charges would affect the prices of virtually all intermediate and final goods.

The General Economic Effects of Carbon Charges

The price changes caused by carbon charges would induce five types of behavior that would tend to reduce emissions of carbon dioxide:

Changes in Consumption Patterns. Carbon charges would raise the prices of energy derived from fossil fuels and of products that require relatively high inputs of energy in their manufacture. Consumers would be induced to substitute other goods and services for those that involve heavy direct or indirect purchases of fossil energy. Fossil fuels are used directly by consumers in automotive gasoline and in natural gas for heating and cooking, and indirectly in items that require energy in their production. Generally speaking, the prices of transportation, plastics, and clay and glassware products would rise relative to the prices of most food and textile products. Consumption of these items would tend to shift accordingly.

Changes in Consumers' Investment Behavior. Consumers would also change their direct consumption of energy by buying more energy-efficient appliances and lighting, by modifying their heating systems, and by making other investments that reduce energy use. All these effects would reduce emissions of carbon dioxide.

Reducing Energy Input in Production Activities. Producers would have an incentive to conserve energy by changing manufacturing processes, by installing more energy-efficient equipment, or by improving housekeeping practices designed to reduce waste. These changes could take the form of substituting other inputs for energy--capital equipment, labor, or materials.

Switching Fuels in Production Activities. Households, businesses, and, especially, electric utilities would be induced to use other fuels. For some this switching would be relatively quick and simple, as in the case of businesses that are equipped to burn more than one type of fuel in their boilers, or of electric utilities that have enough capacity in different types of generating plants to choose among types of fuel to burn. In other cases, fuel substitution would involve investing in new equipment and might require substantial time and expenditure. Fuel substitution can reduce emissions of carbon dioxide because of the differing carbon content of fuels. In particular, substituting oil or gas for coal results in lower carbon emissions for the same amount of energy used. Substituting nonfossil energy for fossil energy would have an even larger effect on emissions of carbon dioxide, but this alternative is likely to be more limited in scope, at least for some time.

Developing New Technologies. Carbon charges would create a financial incentive to develop new technologies that would be more efficient in their energy use. Technological changes are possible in all sectors, including automobiles and other forms of transportation, consumer durables, manufacturing equipment and power plants, and electricity generation. New technologies could also make such noncarbon sources as nuclear, solar, and wind energy affordable, acceptable, and usable on a wide scale.

The revenues from carbon charges, which could be quite large, could be used in a variety of ways. They could be spent to help firms and workers especially hard hit by the new taxes. They could be used to change the structure of public finance by reducing taxes elsewhere, or by retiring federal debt. The Congress could also use new revenues to initiate or expand federal programs, such as reforestation projects, that would reduce carbon dioxide levels.

Not only carbon charges, but other broad-based energy taxes could be employed to encourage cuts in energy use and in emissions of carbon dioxide. The purpose of carbon charges would be to target emissions of carbon dioxide directly; for a given level of revenue generated, carbon charges would be expected to have a greater effect on emissions of carbon dioxide than other energy taxes (see Appendix A for a comparison of carbon charges and an energy tax based on the energy content of fuels).

Carbon charges would also affect the competitiveness of some U.S. industries in international trade--particularly those that use relatively large amounts of energy in their production processes. If major U.S. trading partners were to adopt similar measures, the effects on trade would be lessened.

The adoption of carbon charges by the United States could affect the composition of U.S. trade without significantly changing the overall trade deficit. U.S. industries whose products embodied, directly and indirectly, a large content of carbon-bearing fuels, would be adversely affected, while other less carbon-intensive U.S. industries competing in import or export markets could gain from the carbon charge. The trade deficit exists largely because of macroeconomic factors such as the low level of national saving and the large federal deficit.

Carbon Charges and the Efficient Working of Energy Markets

In principle, carbon charges would have a clear advantage over regulatory approaches to reducing emissions of carbon dioxide. Those advantages are discussed in some detail below. They depend on the efficient working of energy markets. If markets are not efficient--that is, if consumers are led to make choices that are less than optimal from the standpoint of the economy--carbon charges could lose some of their advantage. Impediments to the efficient working of some energy markets include:

o   Lack of information or lack of financial resources necessary for consumers to make efficient choices among consumer durables, such as heating systems or electrical appliances.

o   Situations in which consumers do not directly pay for the energy they consume, as when renters pay fixed charges for utilities as part of their rent.

o   Regulations that distort incentives. Removing controls on oil and natural gas markets has ended many price disparities, but regulation of electric utilities may still cause delivered prices to differ from the real costs of generating electricity and thus bias investment decisions of the utilities in inefficient directions.

o   Difficulties that an individual innovator or developer may have in capturing the economic benefits from creating new technologies.

A number of energy programs, such as utility-sponsored conservation programs and government-supported research and development focused on energy efficiency and alternative forms of energy, address some of these market failures. A carbon charge system could work well with these programs, since it would increase the attractiveness of forms of energy that emit no carbon dioxide, and would also increase the rewards to those who develop and market ways to use alternative forms of energy.

A cost-effective mix of activities or technologies would, by definition, reduce emissions of carbon dioxide with the least sacrifice of economic output. Carbon charges would induce the needed changes in behavior in ways that would approach this least-cost mix. Certainly carbon charges would be more efficient than regulatory measures that would require the government to dictate to consumers or producers what amounts of fossil fuels they should use or what levels of carbon dioxide they should emit.

COMPARING CARBON CHARGES WITH OTHER POLICIES

Imposing a carbon charge would be only one of several ways to reduce emissions of carbon dioxide. Three alternative approaches are available. First, regulations or quotas could be used to restrict the supplies of fossil fuels entering the economy and, consequently, the amount of carbon dioxide entering the atmosphere. Second, demand for energy could be reduced by requiring intermediate and final users to reduce their energy consumption, again most likely through regulation. Third, demand for fossil fuels could be reduced by incentives other than carbon charges. Each of these approaches would have advantages and disadvantages.

Restricting Fuel Supplies by Regulation

Restricting fuel supplies by setting limits on extracting fossil fuels and imposing quotas on imports could directly reduce emissions of carbon dioxide. These regulations and quotas would not be difficult to enforce because of the relatively small number of transactions that take place at the point of primary supply. Limiting supplies would raise energy prices and generate windfall gains for the owners of fossil fuel resources. (The possible magnitude of such gains can be seen from the billions of dollars that flowed from the United States to oil-producing countries when supply was restricted in the 1970s.) Because of their distributional consequences, restrictive supply policies may be neither equitable nor politically feasible.

Production and import quotas would raise prices for users of fossil fuels much as would carbon charges. In the case of carbon charges, however, the difference between the price paid by the purchaser of primary energy and the price received by the seller (the carbon charge) would become government revenue. In other words, carbon charges would increase public income rather than the private income of owners of fossil fuels. The government could capture some of this private income by selling the quota rights to suppliers of fossil fuels. This is a key difference between the oil price shocks experienced in the 1970s and the price increases that would come with carbon charges.[2]

Reducing Energy Demand by Regulation

Causing final users of fossil fuels to reduce their consumption by requiring technological improvements in the current or future capital stock of houses, factories, and vehicles would also reduce emissions of carbon dioxide. At present, technologies that improve energy efficiency exist and could be used more extensively. However, it is impossible to determine the best, most cost-effective set of regulations in all end-use energy markets. An end-use strategy that was effective from the standpoint of cutting energy use might not be the most economically efficient way to reduce energy demands.

Successful demand-side conservation efforts could also have countervailing effects. Any successful demand reduction strategy would be accompanied by downward pressure on market prices, such as was witnessed during the mid-1980s. Lower energy prices would diminish private incentives to conserve energy, so that additional regulations might be needed to meet energy reduction goals. Low energy prices might also erode political support for additional efficiency measures, since the cost savings would be correspondingly small.

---

2. See Brian D. Wright, "The Cost of Tax Induced Energy Conservation, *Bell Journal of Economics*, vol. 11, no. 1 (Spring 1980) The unanticipated and rapid price increases of the oil embargoes contributed to the costs associated with adjusting to lower energy use. But the adverse impact on standards of living was largely a result of the income loss associated with the rising bill for imported oil, not the costs of conservation measures.

Reducing Energy Demand by Incentives Other Than Carbon Charges

Economic incentives other than carbon charges could be used to reduce emissions of carbon dioxide. If properly designed and operated, such incentives could, like a carbon charge, reduce emissions in an efficient way.

One alternative to carbon charges would be a system of tradable carbon permits. The permits would represent the right to produce or import a fossil fuel that would ultimately emit carbon dioxide into the atmosphere. Permits would initially be granted or sold by governments to industries engaged in fossil fuel extraction and importation. Subsequent trading of permits among producers of fossil fuels would, if efficient markets developed, lead to a relatively low-cost reduction in total emissions. Prices of fossil fuels would reflect the value or cost of the carbon permits, and, much as with carbon charges, users of fossil fuels would adjust their consumption in response to the higher prices.

Subsidies to encourage less energy use or less carbon-intensive energy use would be another form of economic incentive. Subsidies would probably be much more targeted toward particular energy uses than would a broad-based carbon tax.

THE BASIS OF THE CARBON CHARGE

Carbon charges would be collected where fossil fuels initially enter the economy--for coal at the mine, for crude oil at the well or dock, and for natural gas at the wellhead. This analysis assumes that all energy imports would be taxed and that taxes would not be rebated on exports, but that no other imported goods would be taxed. Taxing the carbon in fossil fuels would be essentially the same as taxing emissions of carbon dioxide, since emissions are a direct function of the amount and type of fuel burned. The efficiency of combustion and, correspondingly, the amount of carbon dioxide released per physical unit of the fuel, can vary somewhat from one use to another.

Such a carbon charge would be relatively easy to administer because the number of transactions at the point where energy first enters the economy is small. Taxing the first commercial transaction of fossil fuels would be equivalent to a production and import fee. All subsequent transactions in downstream markets would be affected by the charge. Since many primary energy resources are later transformed into other energy forms, such as electricity or motor gasoline, all potential carbon sources would be subject to the charge without risking double taxation.

The carbon charge examined in this study is relatively simple, and many refinements could be made that could affect the efficiency and equity of the charge. Examples of such refinements are:

o   *Exempting or rebating collected taxes for nonfuel uses of fossil fuels.* A relatively small proportion of fossil fuels used in the United States is for purposes other than combustion. These uses do not cause emissions of carbon dioxide--for example, the use of petroleum in production of industrial chemicals and plastics, asphalt, and tires. In the analysis reported on later in this study, fossil fuels were subject to the carbon tax without regard to their final use. Noncombustion uses could be exempted from the tax with relatively minor effects on the results of the analysis.

o   *Taxing the carbon content of goods imported into the United States.* Taxing the fuels used in producing domestic goods would raise their prices relative to imported goods. The drop in U.S. emissions of carbon dioxide might be partly offset by greater emissions elsewhere in the world if imports of energy-intensive goods rose. In principle, imports could be taxed according to the amount of carbon dioxide released in their manufacture. Such taxes would be difficult to set, and would add greatly to the burden of administering the tax system. Such a tax on imports was not assumed in this study.

Not all transactions at the point of extraction or importation of fossil fuels represent an exchange of ownership. Some oil companies, for example, explore, drill, and transport crude oil, refine oil into motor fuels, and sell gasoline directly to consumers. Some electric utilities own coal mines. Additional efforts might be needed to administer the

tax adequately and fairly in cases where internal or subsidiary transactions occur in primary energy.

The carbon charge would be a specific excise tax, not an ad
valorem excise tax (a constant percentage of price). The charge on each
fuel would thus be independent of market prices, and this would be
necessary if the charges were to maintain their relationships to the
carbon content of the fuels. Since the prices of fossil fuels change over
time, ad valorem taxes would not consistently reflect the relative
carbon contents.

THE LEVEL OF THE CARBON CHARGE AND ESTIMATED FEDERAL REVENUES

The larger the carbon charge, the greater would be its effect on
emissions of carbon dioxide and the greater would be the revenue
generated. However, the larger the charge--and the more abruptly it
was imposed--the greater would be the resulting economic dislocation
and the greater would be the loss in output. The actual choice of the
level of the charge would have to take account of these factors.

The Level of the Carbon Charge

Most of the analytic results reported in this study assume that a charge
would be gradually phased in, beginning at a rate of $10 per ton in
1991, and rising by $10 each year until it reached $100 per ton in 2000
(in 1988 dollars). The nominal value of the charge in 1991 is projected
at about $11.30, rising to nearly $160 per ton in 2000, based on
projections of inflation over this period. The study also examines
several variants to this level of charge.

A $100 charge was selected because preliminary analysis showed
that a charge of this order of magnitude would be needed to cause U.S.
emissions of carbon dioxide to return to 1988 levels by the year 2000.
Adopting a more ambitious goal would require a larger charge.

Emissions of carbon dioxide in 1988 were about 5.7 billion tons. The goal of stabilizing them would require emissions to be reduced fully 12 percent below the levels projected in the baseline for 2000.

When fully phased in, the carbon charge of $100 per ton translates into $60.50 per ton of coal, $12.99 per barrel of oil produced or imported, and $1.63 per thousand cubic feet of natural gas (all figures expressed in 1988 dollars). The derivation of these taxes in shown in Table 2. The carbon charge is shown as a percentage of the average projected price of the various fuels at the point at which the charge would be levied.

TABLE 2.   CONVERSION OF A CARBON CHARGE OF $100 PER TON TO CORRESPONDING AMOUNTS PER PHYSICAL UNIT OF OIL, NATURAL GAS, AND COAL

	Product and Physical Unit		
	Oil (Barrel)	Natural Gas (1,000 Cubic feet)	Coal (Short ton)
Average Tons of Carbon per Unit of Fuel	0.1299	0.0163	0.605
Average Carbon Charge per Unit of Fuel (1988 dollars)	12.99	1.63	60.50
Projected Price, 2000 (1988 dollars)	26.66[a]	3.10[b]	23.66[c]
Carbon Charge as a Percentage of Average Price	49	53	256

SOURCES: Congressional Budget Office estimates. Fuel price projections from Energy Information Administration, *Annual Energy Outlook 1990*.

a.   Cost of imported crude oil to U.S. refiners.

b.   Domestic wellhead price.

c.   Domestic mine-mouth price.

The charge on a barrel of oil would be nearly 50 percent of the projected world price in the year 2000. Adding the charge to the projected price of oil would still keep crude oil prices in real terms below the peak level of $39 per barrel registered in early 1981.

Final users of petroleum products would face smaller percentage increases in the prices of products they purchased. Even if the tax was fully passed on to consumers, percentage price increases would be smaller because refining and transportation costs would tend to remain stable. For example, the $100 per ton carbon charge is roughly equivalent to a tax of $0.30 per gallon on gasoline. If gasoline prices were $1.26 per gallon (in 1988 dollars) in 2000, as projected by the Department of Energy, the tax would be about 25 percent of the projected price.

The charge on natural gas would be about 53 percent of projected prices at the wellhead. But the charge would be only about 24 percent of the projected price of gas delivered to residential consumers, and 43 percent of the projected price of natural gas delivered to electric utilities.

The projected mine-mouth price of coal in 2000 is about $24 per ton, with prices for coal delivered to electric utilities projected to average about $33 per ton (in 1988 dollars). By adding $60.50 per ton of coal, the $100 per ton carbon charge would roughly triple the delivered price of coal, reflecting both the high carbon content of coal and its low price relative to other fossil fuels.

Budgetary Effects of a Carbon Charge

Carbon charges could generate significant revenues for the federal government. Additional revenues--that is, the net effect on the federal deficit--from the phased-in carbon charge described above are estimated to be about $13 billion during calendar 1991, the first year of the charge. Revenues would rise as the charge increased annually through 2000. Additional revenues attributed to the carbon charge could reach an estimated $170 billion to $190 billion by 2000, when the tax would be fully phased in. Expressed in 1988 dollars, additional revenues in 2000 could range from $110 billion to $120 billion.

A GENERAL CAUTION

The magnitudes of expected changes in prices and quantities reported in this study are only crude estimates of the outcomes that would be likely to occur in response to carbon charges. Moreover, predicting how production methods and consumption patterns will evolve in the long run is a very difficult task. Small differences in assumptions can lead to vastly different outcomes over long periods of time. This study uses several econometric models that incorporate various elasticity values in order to offer a broad view of likely outcomes.

CHAPTER II

EFFECTS OF A CARBON CHARGE

IN THE FIRST DECADE

Carbon charges would raise the prices of fossil fuels, reducing the use of fossil energy and emissions of carbon dioxide. The use of fossil energy would fall because industries would substitute other, relatively less costly, inputs for fossil fuels. Patterns of consumption would also adjust. The prices of products requiring high levels of use of fossil fuels in their production would rise relative to those with low levels. Consumption would adjust in response to these changing prices. Emissions of carbon dioxide would fall as the use of fossil energy fell in response to these adjustments in production and consumption.

A tax increase of this size would have discernible effects on national income, the general price level, and the balance of trade. Since it would also reduce the federal deficit, the tax increase would temporarily slow the growth of the economy unless the Federal Reserve took action to offset it.

This chapter focuses on the effects of carbon charges over the period between now and the end of the century. Longer-term effects are examined in Chapter III. The analysis was divided into near-term and longer-term for two reasons. First, the effects of carbon charges would be distinctly different during these two periods. In the near term, adjustments in production processes and consumption patterns in response to higher fuel prices would be more difficult or more costly, and hence less likely to occur, than over the longer term. Second, the degree and sources of uncertainty, though substantial even in the near term, are different in the longer term. In the longer term, factors such as the rate of exhaustion of fossil fuel reserves and the rate of introduction of new energy-related technologies must be taken into account.

Between now and the end of the decade, reduced emissions of carbon dioxide would probably have no measurable effects on global warming, though reducing combustion of fossil fuels might have other benefits. Costs would be incurred, however, as the economy adjusted to the higher prices of fossil fuels caused by the carbon charges.

This chapter initially considers the policy option described earlier, namely:

o   A carbon charge beginning in 1991 at $10 per ton, rising annually to $100 per ton in 2000. These amounts are expressed in 1988 dollars. When expressed in current dollars, assuming an annual rate of increase in prices over this period of 3.8 percent, the nominal charge in 1991 would be about $11.30 per ton and the nominal charge in 2000 would be about $160 per ton.

o   The charge would be levied at the point where carbon-based fuels enter the economy--at the mine mouth, wellhead, or dock.

o   The charge would be applied in the United States only. Imports of fossil fuels would be taxed, but imported intermediate and final goods that use fossil fuels in their production would not be taxed.

Three models of the U.S. economy were used to examine the near-term effects of the carbon charge on use of fossil fuels, on emissions of carbon dioxide, and on the overall performance of the economy. These models differ from each other in important ways--one being the degree of emphasis placed on energy markets as against the aggregate performance of the economy, and another being the ease of substitution assumed to exist among fuels and between fuels and other inputs in the production of goods and services. Three models were used rather than one in order to provide a range of estimates, and also because the strengths of each model allow different effects of carbon charges to be examined in detail. The analyses using these models all assume essentially the same baseline levels of fossil fuel use through the year 2000.

The first of these models is the PCAEO simulation model developed by the Energy Information Administration of the Department of Energy. The strength of this model is its detailed representation of

U.S. energy markets. Its main weakness from the perspective of this study is its relatively simple way of linking changes in energy markets with the overall performance of the economy.

The second model is a quarterly econometric model of the U.S. economy developed by Data Resources Incorporated (DRI). This model represents energy markets in a more rudimentary way than does the PCAEO model, but has a much richer structure for examining the effects of taxes on the economy.

The third model, the Dynamic General Equilibrium Model (DGEM), originally developed by Dale Jorgenson and his associates at Harvard University, is used to explore the adjustments in consumption and production that would take place within the economy if those adjustments could be accomplished relatively easily. In the DGEM, a variety of adjustments can occur, and relatively few resources are spent in making the transition toward a less carbon-intensive economy.

CHANGES IN THE EMISSION OF CARBON DIOXIDE AND IN THE USE OF FOSSIL FUELS IN THE NEAR TERM

The models show large differences in their estimates of the effects of a charge of $100 per ton on emissions of carbon dioxide (see Table 3). The PCAEO model shows the least effect, with emissions 8 percent below baseline levels by 2000 but 5 percent above 1988 levels. The DRI model shows emissions 16 percent below the baseline by 2000 (6 percent below 1988 levels), and the DGEM shows emissions 36 percent below the baseline and 27 percent below 1988 levels by 2000.

The differences among the models stem mostly from two factors. The first is the degree of adjustment assumed to take place in response to price changes brought about by the carbon charge. The second is the extent to which the carbon charge affects overall economic activity. A decline in aggregate output following imposition of a carbon charge has secondary and potentially important effects on energy demand and total emissions.

o   The PCAEO model, which shows the least effect of a carbon
     charge on emissions, allows relatively limited possibilities
     for substitution among fuels and between fossil fuels and
     other inputs in production processes. Also, in this model, the
     effects of curtailed aggregate economic activity on emissions
     of carbon dioxide are limited.

o   The DRI model, which shows greater effects on emissions
     than PCAEO but less than DGEM, also allows relatively
     limited possibilities of substitution in production processes.
     Most of the substitution that occurs in the simulation is the
     result of a shift in final demand away from goods heavy in
     energy content toward those with less energy content. The
     DRI model allows the carbon charge to have a greater effect
     on aggregate economic activity than does either of the other
     models.

o   The DGEM model shows the greatest effects of the carbon
     charge on emissions of carbon dioxide. It assumes that the
     economy has great flexibility in adjusting to higher energy
     prices, and that new energy technologies would be readily
     available that are not based on carbon-bearing fuels. The

TABLE 3.   ESTIMATED EFFECTS OF A CARBON CHARGE OF $100
            PER TON ON U.S. EMISSIONS OF CARBON DIOXIDE IN 2000

	PCAEO	DRI	DGEM
Percentage Change in Emissions Compared with 1988 Levels	5	-6	-27
Percentage Change in Emissions Compared with Baseline Levels	-8	-16	-36

SOURCE: Congressional Budget Office.

NOTES: The PCAEO is a simulation model developed by the Energy Information Administration of the
          Department of Energy. The DRI is a quarterly econometric model of the U.S. economy
          developed by Data Resources Incorporated. The DGEM is a general equilibrium model
          developed by Dale Jorgenson and associates.

DGEM ignores the costs of the transition period following imposition of the tax, and, in this sense, represents a longer-term view.

Effects of Carbon Charges on the Demand for Fuels

The effects that carbon charges would have on the use of fossil fuels and on emissions of carbon dioxide in the short run would depend on the reactions of users (in substituting other fuels and in reducing their consumption of total energy) and on the reactions of final consumers of goods and services (in shifting their consumption away from goods and services heavy in energy content).

Simulations using the PCAEO model provide good illustrations of how energy use might respond to a carbon charge. The absolute reductions in fossil energy use estimated with the other two models were greater than those estimated using the PCAEO model, consistent with the greater reductions in emissions of carbon dioxide estimated by the DRI and DGEM models. The PCAEO results are discussed here because of the greater levels of detail about energy markets and energy uses included in this model.

The changes in the use of fossil fuels relative to baseline levels for the year 2000 estimated using the PCAEO model are shown in Table 4. The use of all types of fuels drops with the introduction of carbon charges. The reduction in use of coal is the greatest, and the reduction in use of natural gas is the least. To some extent there is a substitution of gas for coal, principally in the industrial sector and in the generation of electricity. This substitution stems from the greater increase in the price of coal, primarily because of the higher carbon content of coal. But the substitution in favor of natural gas is dominated by a drop in the demand for all fossil fuels since these prices rise relative to baseline projections.

In the PCAEO model, the users of primary forms of energy are divided into the electric utility sector and four end-use sectors--residential, commercial, industrial, and transportation. The end-use sectors use fuels or refined products, such as gasoline, directly, and they also consume energy in the form of electricity.

The relative changes in demand for petroleum, natural gas, coal, and electricity are illustrated in Figure 2. This figure also shows changes in fuel use by electric utilities, and the total changes estimated to occur for the three types of primary energy taken together.

The results shown are for the final year of the 10-year phase-in period when the charge reaches $100 per ton in 1988 dollars ($158 per ton in 2000 dollars). Other adjustments in production processes and consumption patterns would take place over a longer term.

Demand in all the end-use sectors (residential, commercial, industrial, and transportation) falls compared with the baseline when carbon charges are applied, as shown in Figure 2. Coal demand drops substantially in the industrial sector, the only one in which noticeable quantities of coal are consumed, while demand for oil and electricity drops in all sectors.

Fuel Use by Electric Utilities. The electric utility sector accounts for about 80 percent of U.S. coal consumption, and coal-fired plants

TABLE 4.   ESTIMATED EFFECTS OF A CARBON CHARGE OF $100
            PER TON ON PRICES AND USE OF FOSSIL FUELS IN
            THE UNITED STATES IN 2000 (Percentage changes from
            baseline levels)

	Oil	Natural Gas	Coal	All Energy
Prices	21	16	161	n.a.
Use	-6	-4	-13	-7

SOURCE: Congressional Budget Office simulations using the PCAEO model developed by the Energy
          Information Administration of the Department of Energy.

NOTE: n.a. = not available.

generate over half of the total electricity consumed. The response of electric utilities is critical to the effects of carbon charge policies, since coal emits more carbon than other fossil fuels.

In the simulations using the PCAEO model, both overall generation and fuel shares are affected by the carbon charges. Electricity consumption drops by 8 percent in response to higher rates, and utilities reduce their consumption of all fuels. Coal consumption falls by the largest amount, 2 quadrillion Btus, while natural gas and fuel oil con-

Figure 2.
Effects of a $100 Carbon Charge on Use of Various
Sources of Energy in the United States in 2000, by Sector,
as Estimated by the PCAEO Model

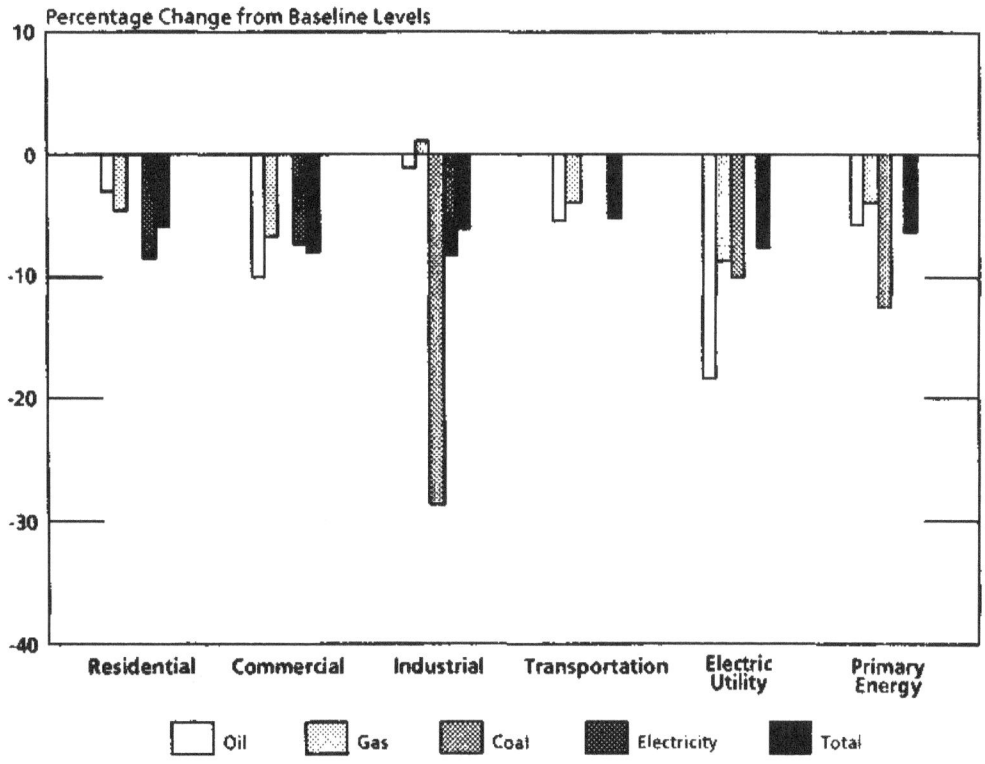

SOURCE: Congressional Budget Office simulation of a $100 carbon charge with the PCAEO model developed by the Energy Information Administration of the Department of Energy.

sumption each fall by about 0.5 quadrillion Btus. These changes combine two effects. To some extent, electric utilities would be expected to substitute natural gas for coal--by cutting back the hours that they run coal-fired power plants, and by increasing their use of gas-fired power plants. These dispatching decisions are made on the basis of the relative costs of operating different units, largely determined by fuel costs. Carbon charges could make some coal plants more expensive to operate than gas-fired plants within the same power pool. Over a 10-year period, however, electric utilities will need to add considerable capacity to meet projected demand. In the base case, relatively more of this added capacity is gas-fired than coal-fired, because there is insufficient time to begin construction of enough additional coal units--beyond those already planned--to meet projected demand. As a result, when projected demand does not materialize because of carbon charges, a large drop in additions of gas-fired capacity occurs, as well as a somewhat smaller drop in additions of coal-fired capacity. If the base case included larger additions to coal capacity between 1991 and 2000, the impact of carbon charges on utility coal consumption would be correspondingly larger.

Demand for energy from other nonfossil sources--geothermal, solar, wind, waste, and others--might expand rapidly if carbon charges made energy from fossil sources sufficiently expensive. However, rapid expansion of the currently very small share of these other nonfossil fuels in generation of electricity--less than 1 percent of total generation--could be expected to satisfy only a small fraction of overall demand for electricity.

Fuel Use in the Transportation Sector. Over 60 percent of petroleum products are consumed in transportation, including gasoline and diesel fuels for automobiles, trucks, and railroad locomotives, and jet fuel for aviation. Oil is the only significant source of energy used in this sector. The decline in energy use in this sector caused by imposing carbon charges would come from reduction in the use of petroleum products (see Figure 2).

Some reductions in petroleum use could be achieved by reducing the use of transportation services. Some consumers and firms could also cut their fuel consumption by reducing the use of their least fuel-efficient vehicles. Over the longer term, rising fuel prices would cause consumers and firms to buy vehicles having greater fuel economy, thus encouraging manufacturers to shift production into more efficient models.

Fuel Use in the Residential Sector. The residential sector directly consumes heating oil, natural gas, and electricity, but little coal. The average price of this energy bundle would rise by about 21 percent, reducing the amount of energy used in the residential sector by an estimated 6 percent. Figure 2 displays these effects. In the short run, energy consumption would be decreased primarily through behavioral changes such as adjusting thermostats or limiting hot water use. Some contribution could come from weatherstripping, insulation, or high-efficiency light bulbs.

Fuel Use in Commercial and Industrial Sectors. The commercial and industrial sectors would reduce overall their energy consumption by an estimated 6 percent to 8 percent. In the short run, commercial establishments could improve their "housekeeping" activities to cut energy use. Changes in industrial processes that save fuel could require several years. Compared with electric utility plants, however, some industrial facilities can easily switch the fuels used to produce heat used in production processes. Natural gas could replace nearly 17 million tons of the 83 million tons of coal used by industry (20 percent), while distillate or residual fuel oil could supplant another 15 million tons (18 percent).[1]

Effects of Carbon Charges on Domestic Energy Producers

The coal industry would be hard hit by a carbon charge. Results from the PCAEO model show a drop in coal use--which is essentially the same as the drop in U.S. extraction--of 13 percent from the baseline. The fact that coal is the most carbon-laden fuel, combined with the cost and difficulty of doing without other fuels in the end-use sectors and the relative ease of reducing coal generation of electricity, makes

---

1.   See Energy Information Administration, *Manufacturing Energy Consumption Survey: Fuel Switching, 1985* (1988), Table 7.

reduced consumption of coal a necessary part of any effective program to control emissions of carbon dioxide in the near term. This would be true whether the policy instruments were carbon charges or regulatory restrictions on coal use.

The use of natural gas would be far less affected by carbon charges than that of coal. In the PCAEO simulation, natural gas demand drops by about 4 percent.

The effects of a carbon charge on the domestic oil industry would depend largely on whether the reduction in energy demand had a noticeable effect in depressing world oil prices. The reduction in oil demand would be only about 3 percent of the U.S. total, or less than 1 percent of world demand. So small a change would have little effect on world oil prices. Domestic production, which depends on the price of competing imports, would also be largely unaffected, and the reduction in demand would be borne almost entirely by imports.

Producers of alternative fuels not subject to the carbon charge would be likely to benefit from the policy. This leaves the coal industry as the only significant loser in the short run from the imposition of carbon charges.

AGGREGATE EFFECTS OF CARBON CHARGES ON THE U.S. ECONOMY

Imposing carbon charges could have a profound general effect on the overall economy in addition to their effects on markets for fossil fuels. The size of the economic effect would depend on how large the charge was, how rapidly it was introduced, whether the charge was anticipated, and on the changes in fiscal or monetary policy undertaken to cushion its effect. Carbon charges could cause reductions in gross national product (GNP), which is a comprehensive measure of the value of all goods and services produced during a time period, and also a measure of aggregate income.[2] The distribution of income could also

---

2.   Gross national product is not an accurate measure of well-being, however. Some of the benefits that could arise from preventing or delaying global warming, such as preventing the inundation of low-lying coastal areas, would not explicitly appear in GNP.

change: some people might be made better off, and some worse off, than the average.

Along with their potential for reducing emissions of carbon dioxide, carbon charges have also attracted attention because of their potential for reducing the federal deficit. Reducing the deficit would presumably confer benefits on the economy in the near term by reducing federal borrowing and thus allowing lower interest rates, and in the long term by increasing the rate of national saving. To achieve these benefits without risking an immediate economic slowdown, however, the Federal Reserve Board would have to ease credit, and the financial markets would have to react positively.

Contractionary Effects Versus Inflationary Effects

Carbon charges--like other excise taxes that would put direct upward pressure on prices--could create more difficulties for the economy than would other types of deficit reduction measures. Any measure to reduce the deficit is likely to reduce consumers' real income. In the case of carbon charges, the reduction would occur as prices rose but incomes did not. Carbon charges, like other excise taxes, would have some additional disruptive effects because they would induce businesses and consumers to change their methods of production and patterns of consumption in order to conserve energy. These changes would be costly in themselves. The price increases caused by carbon charges would probably tend to reduce real consumption by slightly more than a corresponding income tax increase, because they would also reduce real wealth. But more important, the price increases would pose a dilemma for the Federal Reserve. The central bank would have to deal with temporarily higher inflation--which, by itself, should cause it to tighten monetary policy--while faced at the same time with a slowing in economic growth, which might otherwise induce it to loosen monetary policy.

When it had fully taken effect, a carbon charge of $100 per ton would directly raise the consumer price index by 2 percent to 3 percent, assuming that the full amount of the tax was passed forward into the prices of energy and other goods. Potentially, even larger price increases could be triggered if higher prices led to increased wage demands and to spiraling cost increases.

How such price increases would affect the economy would depend on the reaction of the Federal Reserve and the private financial markets--specifically, on whether the higher prices were viewed as an increase in inflation requiring a restrictive monetary response, or whether they were seen as a temporary phenomenon that did not change the underlying rate of inflation. In this latter case, the Federal Reserve could avoid much of the short-run restrictive effects of carbon charges by increasing the money supply and allowing real interest rates to fall.

When oil prices rose massively in the 1973-1974 period, and even more in 1979-1980, the Federal Reserve chose to attack the inflationary effects of oil price increases rather than to ease monetary policy in order to offset the real shock to the economy. The central bank could well act in the same way again. The economic effects of carbon charges, however, would differ in one respect from those of the oil price shocks of the 1970s. Massive amounts of income flowed from the United States to the oil-exporting nations during the 1970s. Carbon charges would not involve this direct transfer of real income to foreigners. Private markets, for their part, often seem to treat increases in excise taxes as representing increases in underlying inflation; for example, when excise taxes rose in Japan and Germany in 1989, nominal interest rates also rose. Most macroeconomic models incorporate the assumption that increases in oil prices would have a persistent effect on inflation as workers increased their wage demands in order to offset the loss in real income.

Even if the Federal Reserve chose to accommodate price increases resulting from carbon charges, it is unlikely that monetary policy could offset all of the effects on GNP. Some of the revenues from carbon charges could be used to reduce other taxes or to fund increased spending on related energy programs or in other parts of the federal budget. Such measures could partially offset the effect of tax increases on aggregate demand, but then would do nothing to diminish the upward pressure on interest rates and thus would be unlikely to eliminate the negative short-run effects on GNP of sudden increases in taxes.

Phased-in Charges

The best way to reduce the strain on the economy from carbon charges would be to allow the economy ample time to adjust, either by announcing the charges well in advance or by phasing them in gradually.

The potential economic effects of carbon charges are illustrated in Table 5. The table shows the simulated effects on GNP, prices, and unemployment of a carbon charge of $100 per ton starting at $10 per ton in 1991 and phased in over 10 years.

With the phased-in charge, losses in GNP rise over the decade, reaching a steady 2 percent reduction in the level of GNP by the year 2000 when the charge is fully in effect. Inflation, the annual rate of change in prices, is projected to be about a third of a percentage point higher than it would otherwise have been through the 1990s, reflecting

TABLE 5.   ESTIMATED EFFECTS OF A PHASED-IN CARBON CHARGE OF $100 PER TON ON GNP, UNEMPLOYMENT, AND PRICES

	1991	1993	1995	1997	2000
Real Gross National Product (Percentage change from baseline)	0	-1	-2	-2	-2
Civilian Unemployment Rate (Absolute change from baseline)	-0.01	0.19	0.48	0.52	0.15
GNP Implicit Price Deflator (Percentage change from baseline)	0	1	2	2	1

SOURCE: Congressional Budget Office simulations using a quarterly econometric model of the U.S. economy developed by Data Resources, Incorporated.

annual price changes while the tax is being phased in. Once the tax is fully in place, most additional inflationary pressure ends, leaving prices about 1 percent higher than they would have been without the tax.

These results assume little action by the Federal Reserve to ease the effect of price increases resulting from carbon charges. Although the estimates assume that one-half the revenues are used for purposes other than deficit reduction, analysis with the DRI model suggests that most of the effect would be felt whether or not the full revenues were applied to deficit reduction. Losses in GNP are only slightly larger when no offsetting changes in other tax revenues or expenditures are assumed.

The transition could be much more difficult if carbon charges were imposed suddenly. Other simulations done with the DRI model suggest that if a $100 carbon charge was introduced in full in 1991 and if monetary policy was not significantly eased, by 1992 the economy could enter a period of stagnation for several years, with little GNP growth and sharp rises in unemployment. These calculations suggest that it would be difficult to mitigate these effects through other measures; GNP would still be sharply reduced even if the full amount of the carbon charges was returned to the economy through reductions in other taxes. Businesses and consumers would find that they must make do with less energy, and they would need time to rearrange their purchasing patterns. Moreover, less energy-intensive methods of production might turn out to be more costly, so economic growth would suffer. Markets would take some time to get back into balance, and afterward levels of output might be lower than they otherwise would have been.[3]

Largely because gradual introduction of a carbon charge would reduce the risk of a severe shock to the economy, a phased-in charge has been made the central case examined in this study. Carbon dioxide emissions would be reduced at a slower rate than if the full charge was imposed immediately, but emissions would begin to approach the same

---

3.   See Douglas R. Bohi and W. David Montgomery, *Oil Prices, Energy Security and Import Policy*, (Washington, D.C.: Resources for the Future, 1982).

levels by the end of the 1990s. As seen in Table 3, the DRI model projects that, with a phased-in charge, emission levels would be 16 percent below baseline and 6 percent below 1988 levels by the year 2000.

Effects on International Trade

Though carbon charges would be unlikely to affect the overall U.S. trade deficit indirectly, they might affect the competitiveness of some industries.[4] The overall trade position is determined largely by macroeconomic factors, such as the levels of national saving and investment. However, exports of industries that are relatively heavy users of carbon-based fuels would be reduced accordingly. Other, less carbon-intensive U.S. industries competing in import or export markets could gain from the carbon charges.

The effect of carbon charges on the U.S. position in international trade might be reduced if other countries also adopted carbon charges or imposed restrictions on the use of carbon fuels that similarly increased costs. In that case, the United States might not be at a cost disadvantage in particular markets, and the worldwide competitive effects could be small. All countries would face losses in GNP because of the domestic costs of reducing use of carbon-bearing fuels, but trade patterns would not necessarily be affected.

POSSIBILITIES FOR REDUCING ENERGY USE

Both the DRI and the PCAEO models show relatively limited possibilities for increasing energy efficiency and substituting other goods for energy in the short run. Although simulations with the DRI model show somewhat greater improvements in energy efficiency in response to a carbon charge than those with the PCAEO model, differences in carbon dioxide emissions in the year 2000 stem largely from differences in GNP--with DRI showing greater emission reductions and greater loss in GNP. Both models focus primarily on short-

---

4.   Carbon charges could indirectly improve the U.S. overall trade deficit if they were used to reduce the federal deficit. See Congressional Budget Office, *Policies for Reducing the Current-Account Deficit* (August 1989).

term responses to higher energy prices, and neither is very good at examining how the structure of the economy could change in response to changing energy prices. The DGEM model completes this part of the picture, though it also leaves something out. The DGEM model provides little insight into what the process of transition to an economy adapted to carbon charges might entail, or how long the adjustment might take.

The Behavior of Businesses and Consumers

The DGEM model traces the effects of changes in the cost of different fuels on the behavior of businesses and consumers. Part of that behavior would be to avoid paying carbon charges by reducing the amount of carbon-bearing fuels used in manufacturing. Charges that could not be avoided in this way would be passed on in the form of higher prices. Those businesses that could reduce their use of carbon-bearing fuels would be at a competitive advantage over those that could not, as buyers changed their purchasing patterns to use less of the goods whose prices had increased most. In the DGEM model, the changes work out as follows:

o   Businesses and consumers with the option of using different forms of energy substitute other fuels with less carbon content for coal and, to a lesser extent, for oil;

o   Businesses substitute other inputs for energy by investing in new kinds of plant and equipment and using more labor and materials and less energy;

o   Businesses that buy semimanufactured goods from other industries decrease their purchases from industries that pass on higher carbon charges, and purchase more from industries in which prices have not been driven up so much; and

o   Consumers rearrange their purchases, buying less of goods whose prices have risen most because of carbon charges and more of goods whose prices have risen least.

Some industries will grow and some will decline as a result of carbon charges. While some overall cost to the economy will be incurred, because patterns of use of energy and of the goods produced by means of energy cannot be changed without some loss in productivity or consumer satisfaction, total employment and levels of economic activity will be maintained as resources shift from declining to growing industries. The cost of the transition will depend on the ease with which consumers and businesses can change their consumption patterns, and on how easily and quickly resources can move from losing to gaining industries.

The amount of economic dislocation in sectors exposed to higher fossil fuel prices depends on how easily producers can substitute other goods for energy. When fuel prices rise, costs will increase most in those sectors that are most energy-intensive and where the potential for substituting other fuels is least. Even if markets were perfect and adjusted with complete efficiency, there would be transition costs associated with the speed and extent of the adjustments.

The major use of energy is in the production of goods and services. A tax on primary energy as it enters the economy would lead to a reallocation of inputs away from energy and its complements into energy substitutes. Consumers would also have an incentive to substitute other goods for energy. Since energy is more important in the production of some goods than others, carbon taxes would cause the prices of energy-intensive goods to rise relative to those of other goods, leading to further reallocations of inputs and consumption.

Energy is also an integral part of U.S. foreign trade, and a carbon charge that affected energy imports would affect other trade with the rest of the world. The imposition of a unilateral carbon charge would lower the relative price of energy-intensive imports compared with domestic energy-intensive goods. Therefore, a portion of the domestic reduction in carbon emissions could be offset by an increase in imports of goods whose production process was energy-intensive. A charge on imported goods other than fossil fuels, based on their carbon content, would significantly increase the administrative costs of a carbon

Figure 3.
Effects of a $100 per Ton Carbon Charge on
U.S. Domestic Supply and Imports, Year 2000

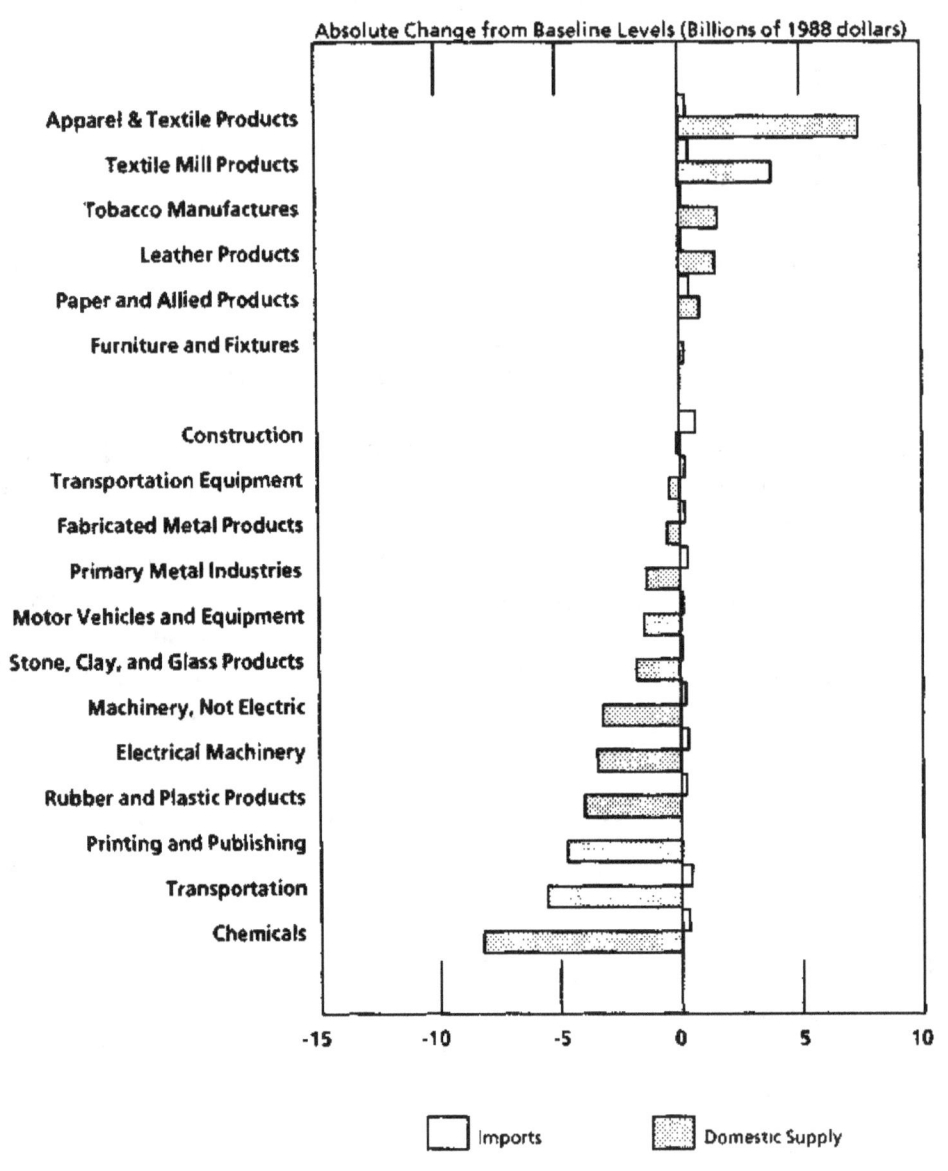

Figure 3.
Continued

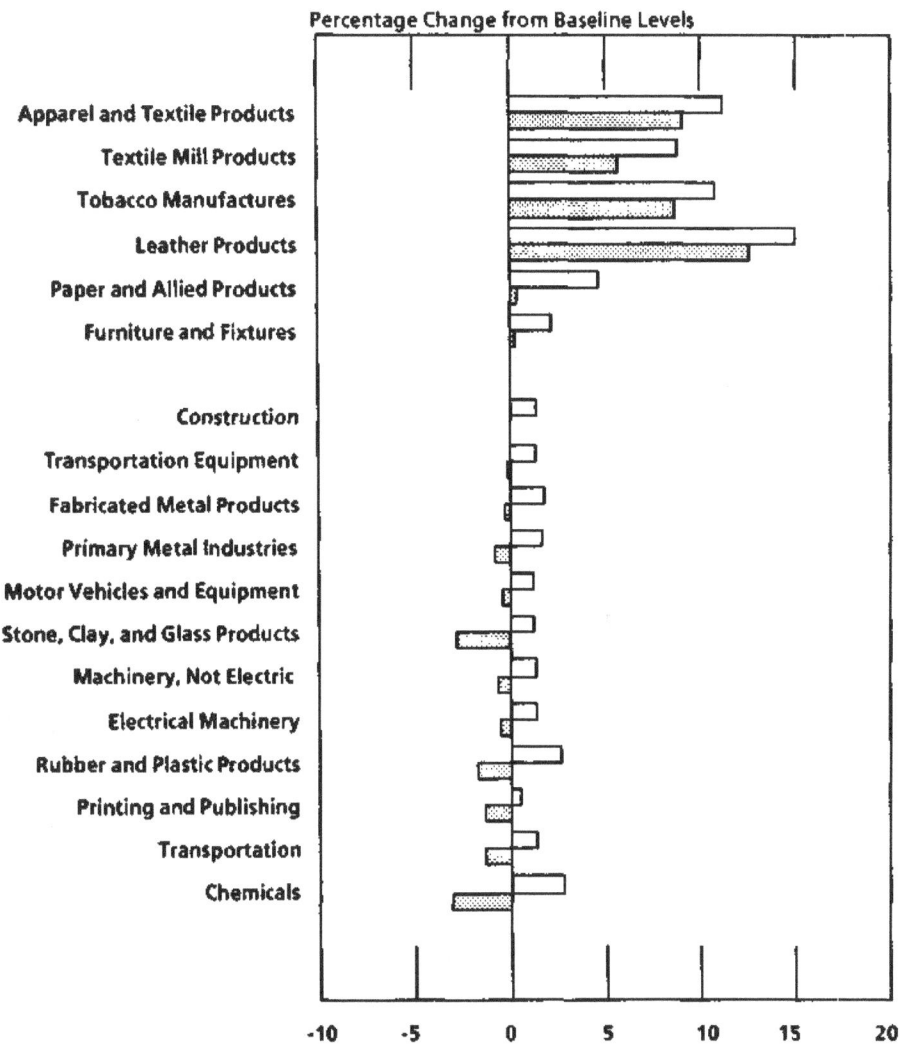

SOURCE: Congressional Budget Office simulation of a $100 carbon charge with the Dynamic General Equilibrium Model developed by Dale Jorgenson and associates.

charge. As the charge would have to be levied on each good according to the fossil fuel used in producing it, the information requirements alone would be extremely burdensome.

This general-equilibrium approach takes account of the various effects of a carbon charge (except for transition costs). Some of the effects, by sector, of a carbon charge simulation in the DGEM model are shown in Figure 3 on page 40. The industries are separated into two groups: non-energy-intensive and energy-intensive. The results demonstrate how consumers would shift purchases away from energy-intensive goods and toward nonenergy-intensive goods. The results also show that as consumption of domestic energy-intensive goods decreased, demand for imports of energy-intensive goods would tend to increase.[5]

The DGEM model assumes that the economy would have great flexibility in adjusting to higher energy prices, and that new non-carbon-bearing energy technologies would be readily available. It ignores the costs involved in moving workers and other resources from losing to gaining industries and occupations. The model represents essentially an extrapolation of past experience with energy price increases. It incorporates all the adjustments in energy markets described above: reduction in GNP, changes in purchasing patterns of consumers, substitution between energy and other factors of production, interfuel substitution, and development of new technologies. Statistical techniques are used to investigate how the changes in each of these areas were in the past related to changes in energy prices. The results of this historical analysis are that a very large response to energy price changes occurred in all these areas.

In forecasting future responses, the model assumes that these relations will continue to hold throughout the forecast period. This assumption is the normal practice in econometric forecasting, and it has much to recommend it. The huge changes in patterns of energy use that occurred in the 1970s and 1980s, and the patterns observed during the many years in which the U.S. economy grew while energy con-

---

5.   DGEM is an open model, so it treats imports as a source of supply competing with domestic output. Thus, imports would be a substitute for domestic supply in meeting purchasers' demands.

sumption fell, are highly instructive. It would be foolish to ignore the lessons of history on the extent to which it is possible to increase energy efficiency and introduce new sources of energy. But one must also recognize how strong an assumption it would be to expect these patterns to continue. Tremendous amounts of waste were wrung out of the economy during the years when rising energy prices forced many energy users to think for the first time about how they used energy. Moreover, the new energy that was substituted for the traditional oil, gas, and coal largely took the form of nuclear power for electricity generation--a trend that seems unlikely to continue.

The DGEM model may thus be biased in a favorable direction so far as the ease of stabilizing carbon emissions is concerned. But the patterns of change shown in the model are interesting, even if its estimates of economic loss are clearly at the lower bound. In particular, the model shows that changes in energy consumption come about largely through readjustments in consumers' purchases toward goods with less direct and indirect energy content and away from those with more. It shows whole industries rising and declining by noticeable amounts, with significant increases in the use of non-carbon-bearing forms of energy. At the same time, the coal industry shrinks to half its current size.

Overall, the DGEM model shows emissions of carbon dioxide being reduced by 36 percent below the baseline. If this occurred by the year 2000, the goal of a 20 percent reduction in emissions would be more than achieved (see Table 6). The cost would be a loss of nearly 1 percent of GNP from what would otherwise be produced in the year 2000, if the entire adjustment process with whatever costs it entailed was complete by then. Total fossil energy consumption would be reduced by more than 30 percent, and the amount of nonfossil energy used would increase by 70 percent.

Changes by Sector

How could such massive changes in energy markets be achieved at so little cost? Reductions would be required in every form of energy consumption in every end-use sector, with the largest percentage decreases occurring in coal consumption (see Figure 4). As end-use consumption of electricity declined relative to the baseline, demand for fossil fuel for electric power generation would also fall, again with the bulk of the reduction concentrated in coal but with no form of fossil energy benefiting as a substitute for coal. (This result reflects the excessive optimism of the DGEM model about the potential supply of nonfossil energy in this time period, but the model shows fairly similar results even if no additional supplies of nonfossil energy are assumed to be available.)

Coal production would fall by 70 percent below the baseline, to less than half the current levels, and even natural gas production would fall by about 15 percent. A decline in U.S. oil demand of about 20

TABLE 6.   EFFECTS OF A CARBON CHARGE OF $100
            PER TON IN 2000 AS ESTIMATED BY THE
            GENERAL EQUILIBRIUM MODEL

	Baseline 2000	$100 per Ton Charge 2000	Percentage Change from Baseline
Carbon Dioxide Emissions (Billions of tons)	6.6	4.2	-36
Energy Consumption (Quadrillions of Btus)			
Fossil energy	82	56	-32
Nonfossil energy	8	13	71
Total	90	69	-23
Real GNP (1988 dollars)	7,137	7,092	-1
Energy Use per Dollar of GNP (1,000 Btus per 1988 dollar)	13	10	-23
Quantity of Labor Services Supplied (Index of hours worked)	1,533	1,538	0.3

SOURCE: Congressional Budget Office simulations using the Dynamic General Equilibrium Model developed by Dale Jorgenson and associates.

percent would be likely to affect world oil prices, so that domestic producers would face lower prices and reduce their production accordingly. The EIA model of world oil markets mentioned above projects that such a decline in U.S. energy demand would cause a drop of $2 to $3 per barrel in world oil prices in the period 2000-2010, and a drop of 300,000 barrels per day in U.S oil production by 2010 (from 7.3 mmbd to 7.0 mmbd). The carbon charge would have amounted to $13 per barrel of oil, so that world market reactions would absorb about 20 percent of the tax.

Figure 4.
Effects of a $100 Carbon Charge on Energy Use in the United States in 2000, by Sector, as Estimated by the DGEM Model

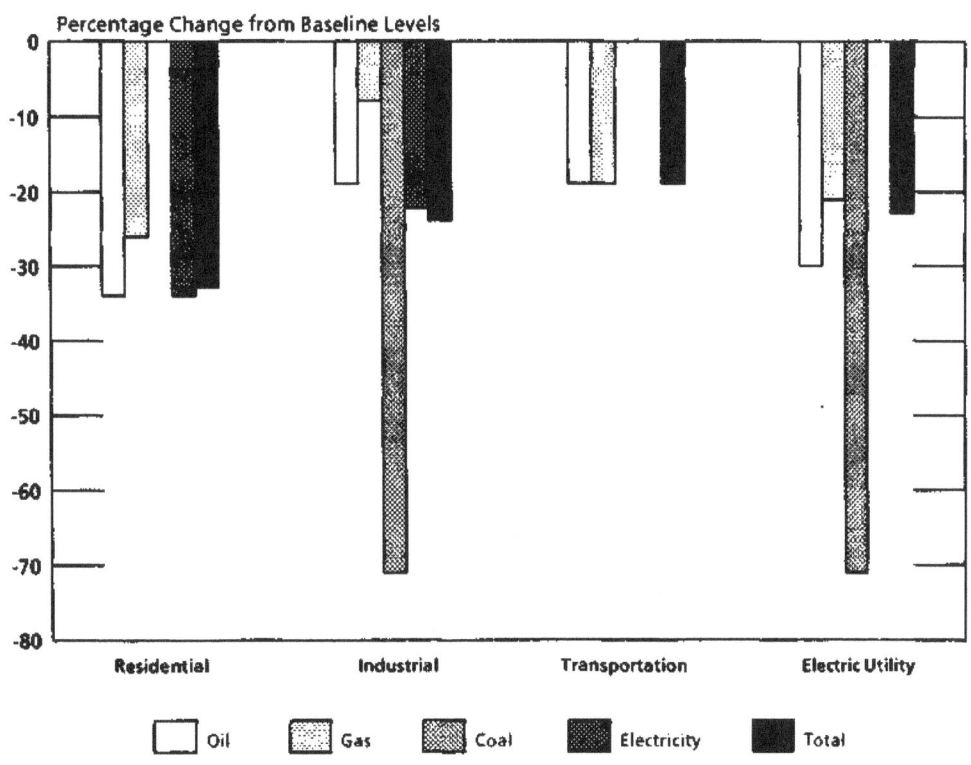

SOURCE: Congressional Budget Office simulation of a $100 carbon charge with the Dynamic General Equilibrium Model developed by Dale Jorgenson and associates.

In the model's residential and transportation sectors, consumers are assumed to be willing to give up 20 percent to 30 percent of their direct energy consumption when faced with prices that are higher by 30 percent or so for oil, gas, and electricity. But they also change their demands for all other goods. It is this change in consumer purchasing habits that allows the substantial reductions in commercial/industrial energy use shown by the model. To some extent, makers of goods also substitute components made with less direct and indirect energy content for those made with more, as shown in Figure 3. Some industries expand by more than 10 percent while others decline in like amount. The amount of energy used per unit of GNP falls by 23 percent according to the DGEM model, compared with a drop of about 6 percent according to the PCAEO model, as should be expected in a model that allows much greater substitution at all levels of the economy. In addition, GNP is sustained to some extent through greater effort, as inputs of labor are substituted for energy. Since full employment is always assumed in the DGEM model, these increased labor inputs represent an additional loss in satisfaction over and above that lost because of lower GNP.

None of the changes described is likely to occur quickly. The immediate effect shown in the DGEM results occurs simply because the model ignores short-term difficulties in adjustment, focusing as it does on long-run responses. Even the year 2000 may be too soon for all the implicit adjustments to work themselves out. Whatever the ultimate ability of the economy to rearrange itself, the kind of economic adjustments projected by DGEM are likely to be accompanied by difficulties and hardships not considered in this model. Moreover, given the likely timing of increases in demand for energy, the exhaustion of gas resources, and the slower development new carbon-free energy sources, maintaining stable levels of carbon emissions beyond the year 2000 might also be very difficult.

CHAPTER III

EFFECTS OF CARBON CHARGES

OVER A LONGER PERIOD

Whatever benefits a reduction in emissions of carbon dioxide would have in averting or delaying global warming, these benefits would not be felt until the next century. The preceding chapter examined the near-term costs of a policy to reduce emissions. This chapter takes a longer view, looking at the effects of carbon charges on carbon dioxide emissions, energy markets, and the economy as far ahead as 2100. Projections extending over 100 years into the future cannot possibly be accurate. But since any response to the threat of global warming would have to be sustained over that long a period of time to be effective, examining the economic effects of carbon charges far into the future cannot be avoided entirely.

Carbon charges that stabilized or reduced emissions of carbon dioxide during the first decade might not be adequate to contain emissions in ensuing decades. Over a longer period of time, the effectiveness of carbon charges would depend on how fast economies grew, how long supplies of fossil fuels lasted, and how rapidly technologies developed that help reduce energy use or that allow nonfossil energy sources to substitute for fossil sources.

The baseline level of emissions--the level assumed to exist in the absence of a policy to stabilize them--is a key factor in estimating the cost of an emissions reduction policy. Demand for energy rises with economic growth. Assuming faster growth would increase the baseline level of emissions of carbon dioxide and raise the tax rate that would be needed to contain the higher rate of emissions growth. At the same time, the exhaustion of oil and gas resources would cause their use to decline during the latter part of the next century, while coal would remain plentiful. Emissions of carbon dioxide would fall as oil and gas use declined, but would rise to the extent that coal was substituted for these other fossil fuels. The rate of development of technologies that reduce the use of energy or that produce energy without releasing carbon dioxide would also have a large influence on emissions of carbon dioxide and the needed size of carbon charges.

A range of possibilities for the twenty-first century is examined in this chapter, which reports on two analyses that differ both in the models they use and, more importantly, in their baseline assumptions. The first analysis was performed by the Congressional Budget Office based on a model originally developed at Oak Ridge National Laboratory by Jae Edmonds and John Reilly. Their original model has since been modified and expanded by the Environmental Protection Agency in its work on global warming.[1] The second analysis was conducted by Alan Manne and Richard Richels.[2] These models address international aspects of the reduction of emissions of carbon dioxide, in that they can project world emissions and be used to examine the global effects of a carbon charge policy. Neither model, however, addresses the question of how changes in the relative prices of energy in different countries would affect international trade flows in nonenergy goods and services.

Analysis using the Edmonds-Reilly model shows that a carbon charge of $100 per ton imposed only in the United States could reduce U.S. emissions of carbon dioxide by more than 20 percent below their current levels by 2100. This amount would entail a reduction of about 4 billion tons out of the 8.2 billion tons of U.S. emissions projected for 2100 in the Edmonds-Reilly baseline.

---

1.   CBO used the version of the Edmonds-Reilly model that was used by the Environmental Protection Agency in its 1989 draft report entitled *Policy Options for Stabilizing Global Climate*, and adopted the same baseline assumptions used by EPA. This version of the model is also referred to as the "Atmospheric Stabilization Framework" or ASF model. Its treatment of energy demand uses conventional econometric models of energy demand, in which demand varies with prices and GNP. Energy supplies are broken down into electricity, oil, gas, and coal; supply depends on price and on the amount of the resource base (for oil, gas, and coal) that remains after cumulative extraction is accounted for. A trend variable representing technological change that reduces the cost of producing each form of energy, and thus increases the amount that can be supplied at any given price, is also included. Oil, gas, and coal are traded in world markets, and the model solves for prices in each year that will equate supply and demand. All energy prices are affected by the price of fossil fuels. Nonfossil fuels are not traded, but are used as inputs to electricity generation. The amount used depends on the relative prices of all fuel inputs.

2.   The results of this model are found in several papers by Manne and Richels, including A.S. Manne and R. G. Richels, "Global CO_2 Emission Reductions--the Impacts of Rising Energy Costs," Stanford University and the Electric Power Research Institute (Palo Alto, Calif., February 1990) and "CO_2 Emission Limits: An Economic Cost Analysis for the USA," *The Energy Journal* (forthcoming), and A.S. Manne, "Global 2100: An Almost Consistent Model of CO_2 Emission Limits," Stanford University (February 1990).

Alternatively, Manne and Richels estimate the size of a carbon charge that would be sufficient to hold U.S. emissions of carbon dioxide in 2100 to a level 20 percent below their 1990 level. But since Manne and Richels project much higher baseline emissions--15 billion tons in 2100--achieving this goal would require a reduction in emissions from baseline levels of over 10 billion tons, or 70 percent of baseline emissions. This larger reduction would mean a larger tax, rising to $400 per ton by 2020 and then declining to about $250 per ton by 2040 and remaining at that level through 2100.

Much of the difference in the conclusions of these two analyses arises from differences in their baseline projections of growth in energy demand and growth in emissions of carbon dioxide. Relatively small differences in key assumptions affecting the baseline can have a dramatic effect over the course of a century. Current understanding makes it difficult to dismiss either alternative as unreasonable. The result is a wide band of uncertainty about what might happen over the next century.

In addition, the two models differ in key assumptions about the availability of carbon-free forms of energy in the early part of the next century. Manne and Richels foresee much greater difficulties in stabilizing emissions between 2010 and 2050 than does the Edmonds-Reilly model. One conclusion on which the models agree is that a unilateral charge sufficient to reduce emissions by 75 percent from the baseline would reduce gross national product by 2 percent to 3 percent from levels it would otherwise achieve during the next century. But in the Manne-Richels model this percentage reduction from the baseline is necessary to achieve the goal of holding emissions to a level 20 percent below current levels, while in the Edmonds-Reilly framework it accomplishes a far more ambitious goal.

THE EDMONDS-REILLY BASELINE AND THE EFFECTS OF CARBON CHARGES

The baseline projections of energy use and emissions of carbon dioxide employed in the analysis based on the Edmonds-Reilly model represent the high-growth scenario, one of two scenarios developed by the

Environmental Protection Agency.[3] In this high-growth scenario, projected emissions of carbon dioxide in the United States reach a peak before 2075 and actually decline slightly thereafter. World emissions, on the other hand, grow almost exponentially (see Figure 5).

Trends Under Current Policies

Stabilization of U.S. emissions in the latter part of the next century follows from four key assumptions made by the Environmental Protection Agency about trends assumed to occur under current policies:

o   The size of the U.S. population is assumed to reach a peak in the middle of the next century and to decline from 2050 onward;

o   U.S. economic growth is assumed to average 1.7 percent per year over the entire period from 2000 to 2100, but to drop to 1 percent a year in the last quarter of the twenty-first century;

o   Even without energy price increases or policy changes, the efficiency of energy use in the United States is assumed to improve at a rate of 1.5 percent to 2 percent per year. Similar improvement is projected in much of the rest of the world, except for the Middle East and the poorest regions in Africa, Latin America, and Southeast Asia; and

o   U.S. emissions of carbon dioxide are assumed to grow more slowly than energy consumption in the first three quarters of the twenty-first century because of a shift out of fossil fuels and into nonfossil energy sources to satisfy energy demand.

Coal use, the most prolific source of emissions of carbon dioxide, is projected to grow at less than 1 percent a year in the United States while it grows at a 2.3 percent annual rate in the rest of the world. Oil

---

3.   In 1986, the Congress requested that the Environmental Protection Agency examine policy options that, if carried out, would stabalize current levels of atmospheric greenhouse gas concentrations. The EPA used a version of the Edmonds-Reilly model to investigate these options in its 1989 draft report entitled *Policy Options for Stabilizing Global Climate.*

and gas use declines worldwide in the latter part of the twenty-first century as these resources are exhausted, but more rapidly in the United States than elsewhere. Nonfossil energy sources in the United States are projected to grow to 2.5 times their level in the year 2000 by the year 2075, and U.S. fossil energy use is projected to be no larger in the year 2100 than in the year 2010. Nevertheless, through 2075, growth in the U.S. economy is sufficient to offset other trends that would reduce fossil energy use, so that U.S. emissions of carbon dioxide are projected to grow, albeit at a declining rate, until the last quarter of the next century.

Figure 5.
Baseline Carbon Dioxide Emissions, Assuming High Economic Growth

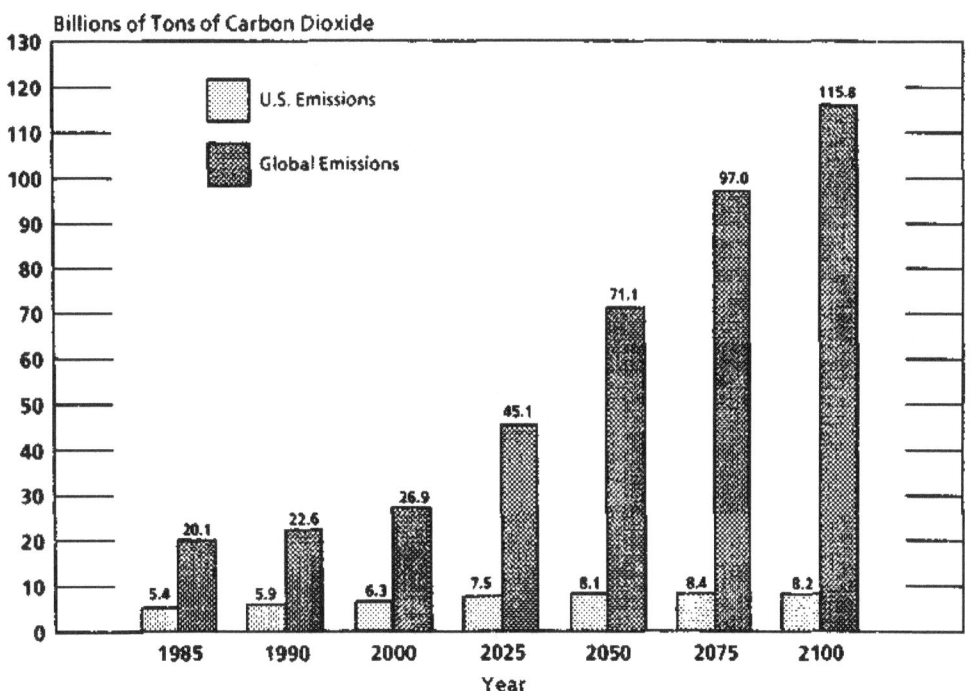

SOURCE: Congressional Budget Office simulations using the Edmonds-Reilly model with the high-
           growth scenario developed by the Environmental Protection Agency.

Energy use and carbon dioxide emissions in the United States actually decline after 2075 as the population begins its assumed decline and as GNP growth slows. Energy use per capita, as well as total energy use, falls in the United States during this period, but grows in the rest of the world. Even so, by 2100 energy use per capita in the rest of the world reaches only one-third of the level in the United States.

Figure 6.
Contributions to Carbon Dioxide Emissions, United States and
Rest of the World, Assuming High Economic Growth

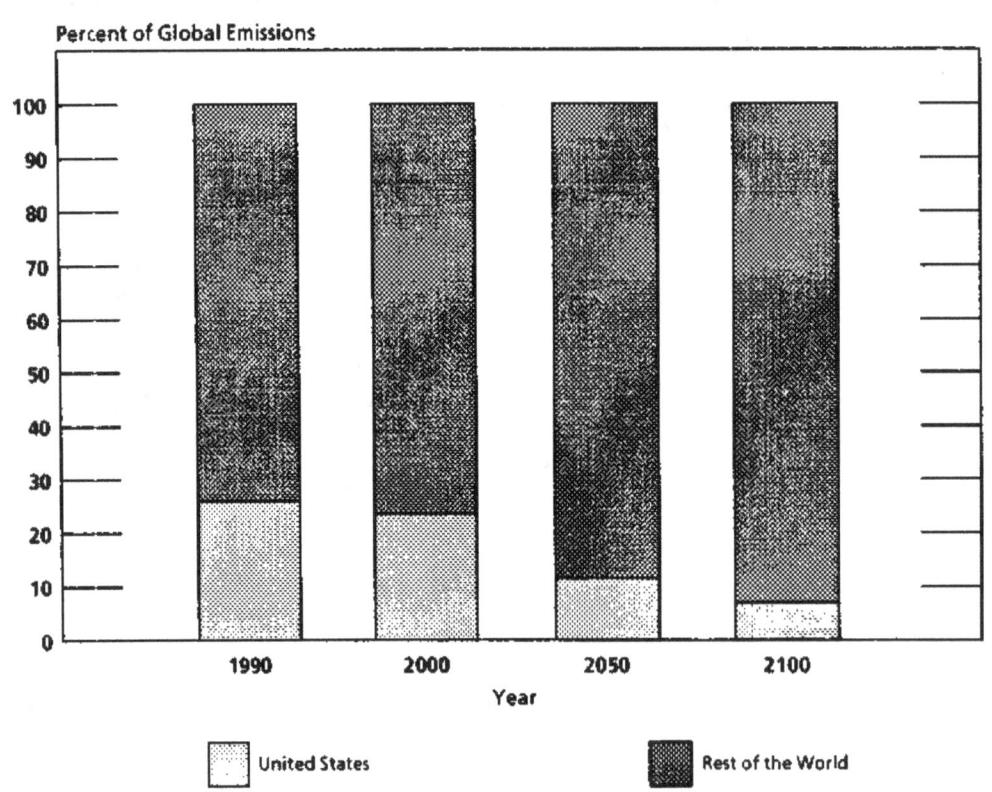

SOURCE: Congressional Budget Office simulations using the Edmonds-Reilly model with the high-
          growth scenario developed by the Environmental Protection Agency.

A major implication of these baseline projections is that the United States is responsible for a falling share of world emissions of carbon dioxide over the next century (see Figure 6). The declining relative contribution of the United States to the global warming problem implies that unilateral actions taken by the United States to reduce emissions of carbon dioxide could have little effect on the world total.

Effects of Imposing Carbon Charges

A unilateral carbon charge of $100 per ton would cause U.S. emissions of carbon dioxide to decline for a few years, and even after growth resumed, emissions of carbon dioxide would barely exceed 1988 levels at their peak. By 2100, U.S. emissions of carbon dioxide would be more than 20 percent below their 1988 levels (see Figure 7). The reduction

Figure 7.
Effects of Carbon Charges on U.S. Emissions of Carbon Dioxide, Assuming High Economic Growth

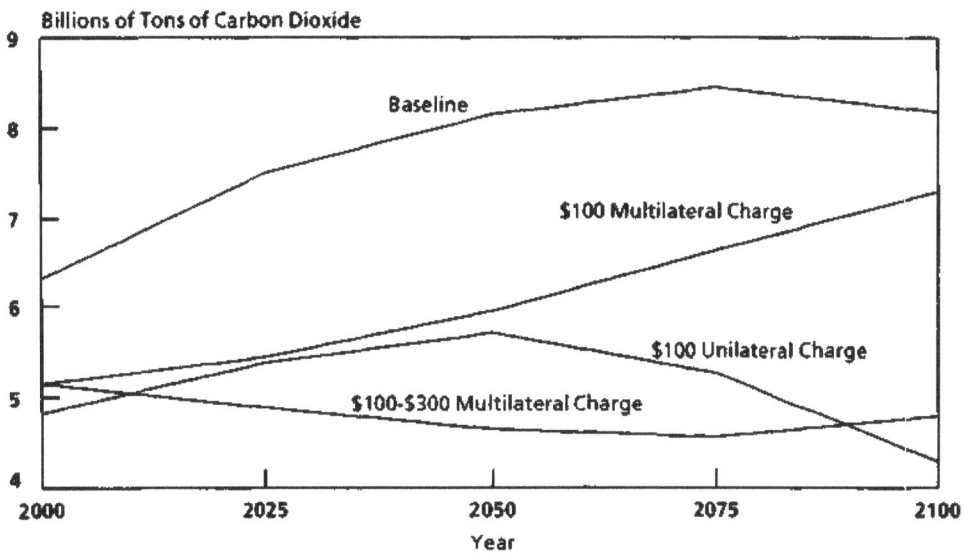

SOURCE: Congressional Budget Office simulations using the Edmonds-Reilly model with the high-growth scenario developed by the Environmental Protection Agency.

in emissions of carbon dioxide caused by the $100 charge would amount to about 4 billion tons per year in 2100, or just under 50 percent of baseline emissions. The cost of a sustained unilateral charge would be roughly 1 percent of GNP throughout the century (see Figure 8).

For purposes of comparison with the Manne and Richels analysis of a charge reaching $250 per ton in 2100, the study also analyzed a more aggressive policy--a unilateral carbon charge rising from $100 to $300 per ton from 2000 to 2100 (not shown in the Figures). In the

Figure 8.
Effects of Carbon Charges on U.S. Gross National Product,
Assuming High Economic Growth

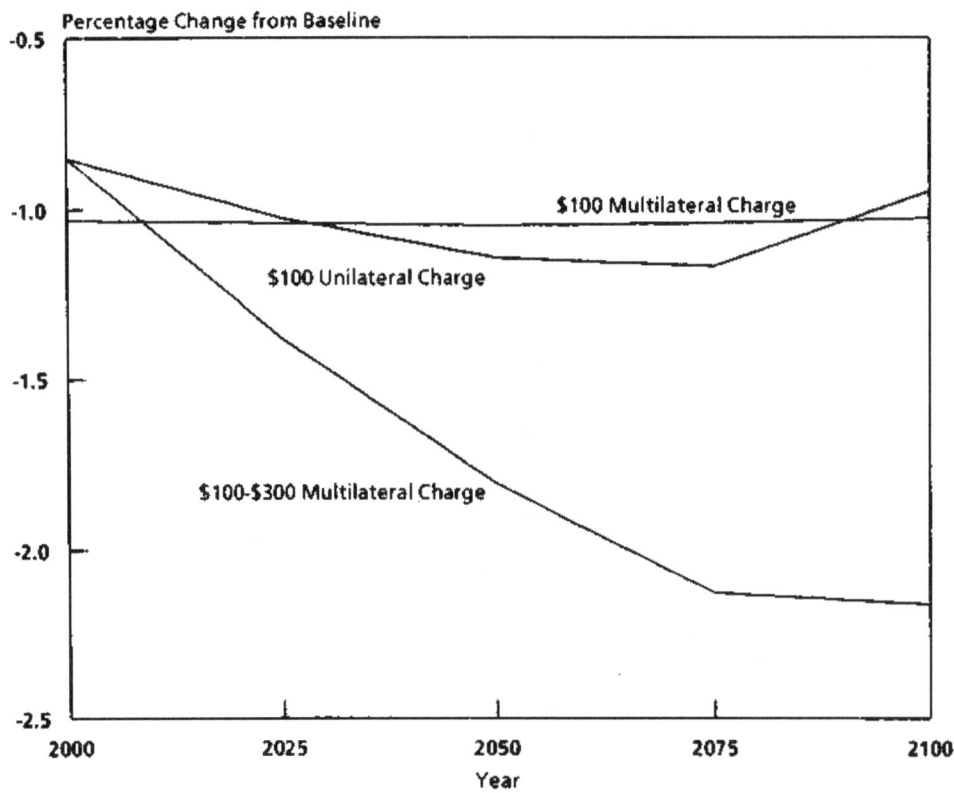

SOURCE: Congressional Budget Office simulations using the Edmonds-Reilly model with the high-
           growth scenario developed by the Environmental Protection Agency.

2

Edmonds-Reilly model such a charge would cause a reduction in emissions to a level 75 percent below the baseline projection, with costs growing to nearly 3 percent of GNP per year by 2100.

The Edmonds-Reilly model can also be used to look at how energy prices and demand in the rest of the world would respond to changes in U.S. energy use. With a unilateral $100 charge per ton, there would be small offsetting reactions from other countries. Reductions in U.S. demand for oil, gas, and coal would lead to reductions in the prices at which these fuels are traded in the rest of the world, and to slight increases in demand in other countries. The result is that worldwide emissions of carbon dioxide (including the United States) would decline by less than the U.S. reduction. Nearly 10 percent of the U.S. reduction would be offset by increases in energy use in other countries.

The Edmonds-Reilly model assumes a single world market for all fuels, so that a reduction in U.S. demand would have a depressing effect on world prices. This would provide some stimulus to energy consumption elsewhere, so that some offsetting increase in carbon emissions could be expected. The increase would be small, however, because by 2100 the United States would account for only a small share of world energy consumption in either the baseline case or the carbon charge case. Thus, the effects on world markets of any change in United States consumption would be limited.

The Edmonds-Reilly model does not examine how imports of energy-intensive goods into the United States might change as the result of unilaterally imposing a carbon charge. If other countries increased their production of energy-intensive goods to satisfy demand in the United States, while production of these goods in the United States declined, U.S. carbon emissions might be even lower and emissions in the rest of the world might be higher.

Unilateral Versus Mutilateral Charges

From a global perspective, unilateral adoption of a carbon charge by the United States would have an insignificant effect on total emissions of carbon dioxide (see Figure 9). A multilateral charge would reduce emissions both worldwide and in the United States by a significant amount relative to the baseline (see Figures 7 and 9). A multilateral charge of $100 per ton would reduce world emissions substantially, but still fall far short of achieving global stabilization. A carbon charge that began at $100 per ton in 2000 and rose to $300 per ton would do more, removing about half of the growth in world emissions of carbon dioxide. Such a charge would also hold U.S. emissions of carbon dioxide to a level about 15 percent below their 1990 level. A multilateral charge would also be less likely to alter patterns of trade in manufactured goods.

Figure 9.
Effects of Carbon Charges on Global Carbon Dioxide Emissions, Assuming High Economic Growth

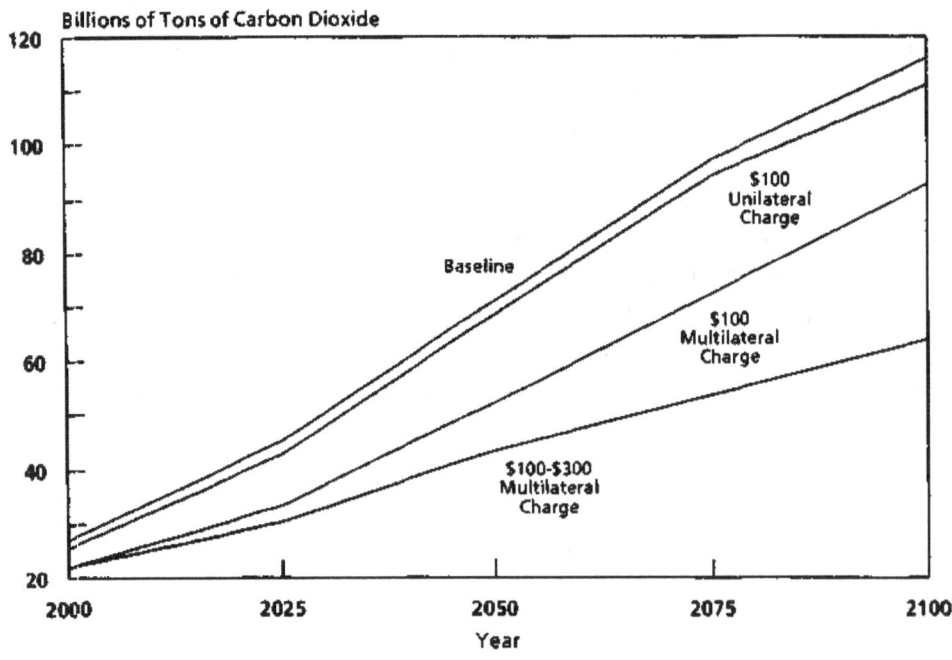

SOURCE: Congressional Budget Office simulations using the Edmonds-Reilly model with the high-growth scenario developed by the Environmental Protection Agency.

The pattern of energy prices and use that emerges when a multilateral charge is assumed differs from that when it is a U.S. action only. As mentioned previously, a unilateral charge produces a dramatic reduction in U.S. coal demand, by raising the price that consumers pay for coal by over 200 percent. But a multilateral charge of $100 per ton allows U.S. energy use and emissions of carbon dioxide to be larger than they would be with a unilateral charge of the same amount.

The reason for these patterns is to be found in the change in pretax world energy prices that results from imposing a carbon charge. When the rest of the world also imposes a carbon charge, coal use throughout the rest of the world falls. This reduction in world coal demand, of which the United States accounts--even in the baseline--for only 7 percent in 2100, causes a 20 percent decline in the price of coal charged before tax by coal suppliers. The pre-tax price of coal decreases because, with a large reduction in the amount of coal required worldwide in the next century, the cost of producing coal drops dramatically. In the baseline, world coal demand grows from about 125 exajoules in the year 2000 to 1,100 exajoules by 2100.[4] To supply this much coal it is necessary to move to coal reserves that are increasingly costly to extract. With the multilateral charge of $100 per ton, world coal demand grows to under 900 exajoules by 2100. This much coal can be produced at a marginal cost that is lower than the cost of producing the amount demanded in the baseline. Thus, with a $100 charge, the mine-mouth price of coal, which is based on the marginal cost of producing coal, declines.

Since coal is assumed to be traded worldwide, this cost reduction would benefit the United States as well. Consequently U.S. coal use would be much higher with a multilateral carbon charge than with a unilateral charge, because the mine-mouth price of coal in the United States would be much less than if carbon charges were imposed in the United States only. The United States would consume more energy, and emit more carbon dioxide, under a multilateral agreement on carbon charges than if it acted unilaterally. But as can be seen in Figure 7, the United States would not have emissions as large, in

---

4.   An exajoule is 10^{18} joules. A joule is a unit of energy equal to 0.948×10^{-3} British thermal units.

either case, as it would in the baseline that assumes no country takes any action to control emissions.

These results suggest that international negotiations on carbon dioxide *stabilization* policies might differ greatly from negotiations on *carbon charge* policies. Agreeing on a multilateral approach to charges would mean that energy prices in the United States would be lower than with a unilateral U.S. charge of the same nominal amount. This would add up to an implicit agreement that the United States could emit more carbon dioxide than if it acted unilaterally to adopt such a charge.

These descriptions of differences between unilateral and multilateral carbon charges also depend on assumptions about how world energy markets will work a century from now--in particular, that prices will be established on flexible and competitive markets and that the costs of producing coal will rise rapidly with increasing cumulative production. World energy markets are likely to become more tightly integrated over the next century, as the volume of energy traded around the world grows. But whether they will be competitive or continue to be under the influence of cartels or major suppliers is impossible to tell. Moreover, costs of production in the United States and other countries cannot be forecast with precision so far in advance. If reductions in U.S. coal demand did not produce as large a drop in world prices as assumed--as would be the case if the United States remained largely self-sufficient in coal and a unified, competitive international coal trade did not develop, then there would be little difference between unilateral and multilateral charges in terms of their effects on carbon dioxide emissions from the United States.

THE MANNE AND RICHELS BASELINE AND THE EFFECTS OF CARBON CHARGES

The Manne and Richels perspective on the likely evolution of U.S. energy demand in the baseline, and on the total cost of stabilizing emissions, is very different than that described above. Manne and Richels assumed a target of zero growth in emissions of carbon dioxide through 2000, a target of a 20 percent reduction by 2020, and zero growth from then on (see Figure 10).

Through the year 2000, their conclusions are similar to those of the Dynamic General Equilibrium Model discussed in Chapter II--that zero growth in emissions of carbon dioxide could be achieved with modest losses in GNP. As in the case of the DGEM, the Manne-Richels approach may be overly optimistic about the ease of adjustment, because it does not consider the costs of transition to lower energy use but concentrates on what would happen beyond 2000. Manne and Richels conclude that the United States would lose about 3 percent of GNP in every year from 2030 on, in order to reduce its emissions to a

Figure 10.
Effects of Carbon Charges on U.S. Carbon Dioxide Emissions
(Manne and Richels)

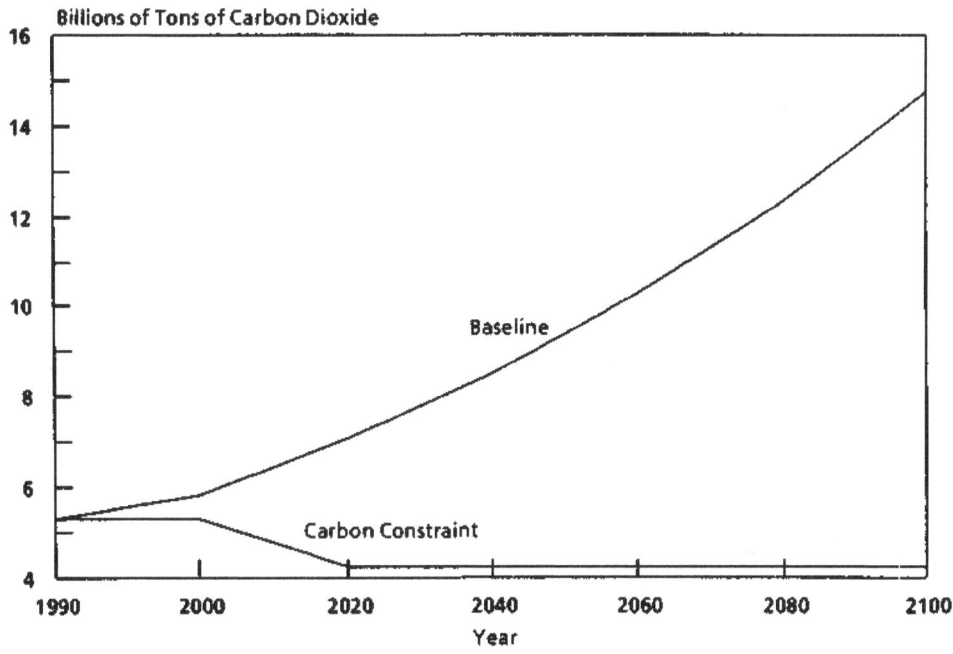

SOURCE: Congressional Budget Office, from Alan S. Manne and Richard G. Richels, "Global CO$_2$ Emission Reductions--the Impacts of Rising Energy Costs" (preliminary draft, February 1990).

level about 20 percent below 1990 levels. This is about the same result as is obtained from the Edmonds-Reilly model with a $100 charge, but it requires a much larger reduction from baseline emissions. As a result, Manne and Richels project a charge that rises to $400 per ton by 2020, and then recedes to about $250 per ton as less expensive non-carbon-bearing technologies finally become available on a wide enough scale.

This peaking of the charge implies that stabilizing emissions might be a particularly difficult goal in the early part of the next century, because of a mismatch of time scales among the developments that make for changes in energy balances. In performing their analysis, Manne and Richels reached some dramatic conclusions about this mismatch of time scales:

o   Long-term growth rates for energy consumption imply a growing need for forms of energy with little or no carbon content through the next century;

o   Domestic natural gas resources may begin to be exhausted in the early twenty-first century;

o   Technologies allowing the commercial use of carbon-free energy sources may not be widely available until after 2030.

By setting a rate of improving energy efficiency in the United States equal to 0.5 percent a year, Manne and Richels attempted to choose a middle ground in their assumptions about the factors that drive growth in energy use. Even higher forecasts of energy demand growth may be justified by recent research into relations between energy prices and economic growth, energy efficiency, and technical change.[5] This research suggests that in the absence of energy price hikes during the 1970s, the economy would eventually have become less energy-efficient--that is, have used more energy per dollar of GNP. The conventional wisdom in energy forecasting has been that the

---

5.   W.W. Hogan and D.W. Jorgensen, "Productivity Trends and the Cost of Reducing CO_2 Emissions," Harvard University, January 1990 (processed).

energy-GNP ratio will decline at a rate of about 1 percent a year, while the recent research suggests that it may be declining only slightly.

Concern about the size of the U.S. natural gas resource has existed for some time. The gas glut that existed through most of the 1980s muted that concern, and any estimate of the size of the natural gas resource is highly speculative. Nevertheless, some evidence indicates that natural gas will become increasingly difficult to find and produce sometime early in the next century if current rates of extraction continue. If this is the case, it will become increasingly difficult to substitute what is now the best available alternative for coal as a means of reducing emissions of carbon dioxide.

No means is now known that could, at acceptable cost, remove and store carbon dioxide after fossil fuel is burned. Finding clean alternatives to coal requires developing energy sources that do not release carbon dioxide into the atmosphere, and that can be used widely enough to supply the entire increment to demand. Some resources, such as geothermal energy, are currently available but cannot be expanded sufficiently. Others, such as fusion energy, are not likely to be available until late in the twenty-first century. Developing inherently safe and economical fission reactors might provide a sufficiently large source. New technologies for improving energy efficiency may also have a role to play. But the time horizons for such alternative sources of energy indicate that it will be difficult to hold the line on emissions of carbon dioxide.

Given the prospect of substantial growth in the demand for energy and the limitations on the supply of cleaner fossil fuels, preventing growth in emissions of carbon dioxide will require finding noncarbon-bearing energy sources that can supply the entire increment to demand in every year. If those sources are not available, the carbon charge must be set high enough to choke off growth in demand if stabilization is to be accomplished. Given the paucity of alternatives, choking off energy demand means giving up a large portion of the economic growth that would otherwise be possible. This bleak prospect lies behind the conclusions of Manne and Richels about the cost of limiting fossil fuel emissions during the first half of the next century.

If sufficient time is allowed to introduce new nonfossil energy sources, the charge required to achieve stabilization of emissions falls.

Manne and Richels find that by the latter part of the twenty-first century a charge of around $250 per ton would be sufficient to ensure that increases in the demand for energy were met from sources with no emissions of carbon dioxide. This conclusion also points out the importance of developing technologies capable of providing enough energy to satisfy all the increases in demand likely to accompany continued economic growth. Carbon charges would induce a substantial amount of research leading to alternative energy sources. Other policy measures, such as direct tax incentives for research and development, could also help spur the development of such technologies.

EFFECTS OF EMISSION REDUCTION ON GLOBAL CARBON DIOXIDE CONCENTRATIONS

The long-run effects of different carbon-charge strategies on concentrations of carbon dioxide in the atmosphere are shown in Figure 11. Since the United States is responsible for about 25 percent of current global emissions of carbon dioxide, and the U.S. share is expected to fall in the future, unilateral action is unlikely to achieve significant reductions in the long run. To make comparison of different strategies easier, analysts sometimes use the time required for a doubling of the concentration of carbon dioxide in the atmosphere. Simulations with the Edmonds-Reilly model suggest that unilateral imposition of a $100 charge per ton would delay the doubling of atmospheric concentrations of carbon dioxide by only three years. The results of multilateral action could be much more dramatic--a delay of about 17 years in the time of doubling. If a rising charge--as high as $300 by the year 2100--was adopted, doubling might be put off until after 2100.

A multilateral agreement on measures to curb global warming could involve commitments to reduce emissions of greenhouse gases and curb deforestation. Negotiations would probably be more compli-

Figure 11.
Effects of Carbon Charges on Global Concentrations of
Carbon Dioxide, Assuming High Economic Growth

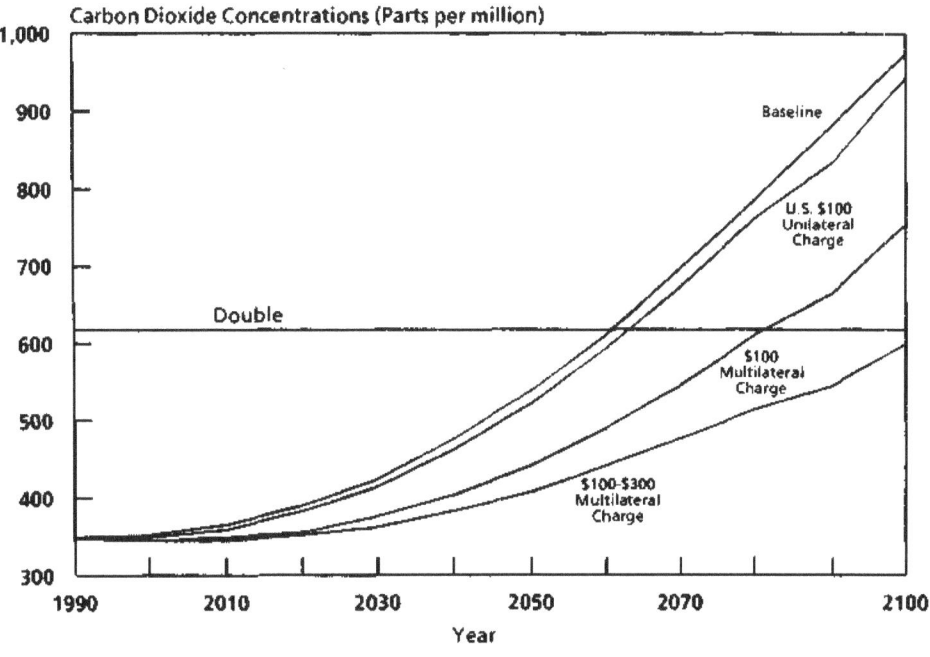

SOURCE: Congressional Budget Office simulations using the Edmonds-Reilly model with the high-
           growth scenario developed by the Environmental Protection Agency.

cated than was the case for the CFC protocol.[6] There are only a few
producers of CFCs, and the same producers are likely to be the leading
manufacturers of the chemicals that substitute for CFCs. Also, the
benefits that would be achieved by action to delay the effects of global
warming would not be uniform among countries, and some countries
(in colder climates) could conceivably gain from a limited degree of
global warming.

---

6.   The Protocol on Substances that Depleted the Ozone Layer was negotiated at a conference spon-
      sored by the United Nations Environment Programme (UNEP). The conference met in Montreal in
      September 1987. The so-called Montreal Protocol established specific obligations to limit and
      reduce the use of chlorofluorocarbons (CFCs). Fifty-six members of the United Nations attended
      the Conference; the United States and twenty-three other nations signed the Protocol. In 1989, at a
      UNEP meeting in Helsinki, eighty countries agreed to ban CFC production and consumption by the
      year 2000. This agreement is more restrictive than the earlier Montreal Protocol, which called for a
      50 percent cut.

The "Group of Seven" nations account for roughly 40 percent of global emissions of carbon dioxide. The four largest contributors of the Group account for about 35 percent: the United States, Japan, West Germany, and Great Britain. As these countries have regular economic summit meetings, a consensus may be feasible within the G7. Except for the United States, all of the countries represented at the July 1990 summit meeting of the Group have pledged to stabilize greenhouse gas emissions by, at least, early in the next century.[7] Since China, the Soviet Union, and some of the larger developing countries are expected to account for an increasing share of global fossil fuel use, they will eventually have to be incorporated in any agreement to reduce emissions of greenhouse gases.

CONCLUSIONS

Carbon charges can work to reduce emissions of carbon dioxide from fossil fuel combustion in the United States. A gradually introduced carbon charge reaching $100 per ton of carbon (in 1988 dollars) by the year 2000 would probably stabilize carbon dioxide emissions in this country through the year 2000. Achieving much greater reductions by the year 2000 would be unlikely without rapid and massive changes in energy production and use. Imposing carbon charges rapidly enough and at a high enough level to achieve reductions in U.S. emissions would risk several years of economic stagnation and high unemployment. These transition costs could be reduced substantially by phasing in the charge and returning some of the revenues to the economy through reductions in other taxes.

Over the longer term, maintaining a carbon charge at $100 per ton would mean keeping GNP at least 1 percent lower than it would be without charges. Even higher charges would be required to prevent growth in emissions of carbon dioxide after 2000. A $100 charge imposed unilaterally by the United States might delay the effects of global warming by only a few years, while a charge of that amount imposed by all countries might buy almost two decades.

---

7.   See *The New York Times* (July 10, 1990).

The best policy for reducing the use of fossil energy would employ economic incentives rather than government regulations. Economic measures, such as carbon charges, would provide measured incentives for energy consumers to reduce their reliance on carbon-bearing fuels in a multitude of ways. Other policies would not address as wide a range of possible responses. Programs to stimulate research and development may also be needed, along with other programs designed to address market imperfections that hinder responses to measures designed to enhance energy efficiency.

The question then becomes whether the costs imposed by carbon charges, or by any other program equally effective in controlling emissions of carbon dioxide, would be worth bearing in order to obtain the likely delay in global warming. The cost of approaching the global warming problem through reducing fossil energy use suggests that there would be merit in looking into other ways of dealing with global warming.

Other greenhouse gases are more reactive--more effective in causing global warming--and potentially cheaper to control than carbon dioxide. Some analyses have suggested that complete elimination of chlorofluorocarbons should be considered before drastic reductions in fossil energy use are undertaken. Methane is another atmospheric trace gas whose control is worth examining. Cultivation of plants and forests that remove carbon dioxide from the atmosphere through photosynthesis could be of obvious benefit. Clearly research and development has a role in discovering ways of improving energy efficiency and finding new technologies for using forms of energy that do not release carbon dioxide.

Finally, it should be emphasized that global warming, if it occurs, will be a long-run problem--covering 50 years and more. Over comparable periods, in the past, the location of economic activity has changed dramatically, new techniques of farming and new kinds of crops have come into use, and the capital stock of the nation has turned over many times. Since the economy will continue to change, more attention might be paid to creating incentives for it to change in ways that will lessen the likelihood of global warming.

COMPARING CARBON CHARGES WITH
OTHER WAYS OF TAXING ENERGY SOURCES

Two ways of taxing energy are shown in Table A-1, which compares carbon charges and Btu taxes designed to collect the same net revenues of $15 billion.[1] A Btu charge, based on the energy content of primary fuels and including other sources such as nuclear energy and hydropower, would differ in two major ways from a carbon charge. First, since the Btu charge would have a base 15 percent broader than that of the carbon charge, it would raise the same revenues with lower average rates on fossil fuels. Second, since coal has more carbon per unit of energy than other fuels, a Btu tax would impose less burden on coal than would a carbon charge. But the differences are matters of degree; as can be seen Table A-1, when measured as percentages of end-user prices, Btu charges and carbon charges are both much larger for coal than for oil and gas. The energy form on which the largest revenues would be collected is in both cases oil, not coal, because oil is the dominant source of energy in the United States. Btu charges would tend to discourage the introduction of forms of energy such as nuclear power and hydropower that do not emit carbon dioxide.

An ad valorem tax on all fuels at the retail level--a third method of taxing energy--would have a much smaller effect on coal, and a much larger effect on oil and gas, than a Btu tax. It would probably have a much smaller effect on the choice of fuels than either of the other taxes, so that its influence on energy consumption would be mostly through conservation. An ad valorem federal tax could be quite difficult to administer because of the large number of points at which energy is purchased by end users, a problem the two other taxes would avoid. To get around this difficulty, some proposals for an ad valorem tax have suggested calculating what an average ad valorem tax on each major type of fuel would be if the tax were collected at the point of final sale,

---

1.   Revenues are net of reduced income and payroll tax revenues.

TABLE A-1.   TWO WAYS OF TAXING ENERGY: CARBON CHARGES
              AND BTU TAXES (Dollar amounts in 1989 dollars)

	Oil	Natural Gas	Coal	Other	Total
Physical Unit	Barrel	1,000 cubic feet	Short Ton	n.a.	n.a.
Total Consumption (Quadrillions of Btus)	34.4	19.2	19	12.8	85.4
End-User Price (Dollars per million Btus)	8.38	5.41	1.47	n.a.	n.a.
Electric Utility Price (Dollars per million Btus)	2.75	2.39	1.47	n.a.	n.a.
Carbon Charge					
Dollars per million Btus	0.28	0.20	0.35	0.00	n.a.
Dollars per unit of fuel	1.63	0.20	7.57	0.00	n.a.
As a percentage of sales price	3-10	4-8	24	n.a.	n.a.
Per ton of carbon	12.51	12.51	12.51	n.a.	n.a.
Revenue (Billions of dollars)	7.23	2.85	4.92	0.00	15.00
Btu Tax					
Dollars per million Btus	0.23	0.23	0.23	0.23	n.a.
Dollars per unit of fuel	1.36	0.24	5.13	n.a.	n.a.
As a percentage of sales price	3-9	4-10	16	n.a.	n.a.
Revenue (Billions of dollars)	6.04	3.37	3.34	2.25	15.00

SOURCE: Congressional Budget Office.

NOTES: Revenue estimates are net of reduced income and payroll tax revenues. The carbon charges
         and Btu taxes are estimated to collect the same net revenues. The charge or tax as a
         percentage of sales price is calculated in the following ways: for coal, as a percentage of the
         average price of coal delivered to electric utilities; for oil, as a range of percentages of the retail
         price of gasoline and of the price of residual fuel oil delivered to electric utilities; and for
         natural gas, as a range of the percentages of the residential price and of the average price of
         gas delivered to electric utilities.

         n.a. = not applicable.

and converting it to a specific levy collected earlier in the distribution chain. Because of the complexity of the contracts and transportation arrangements involved in the sale of coal and natural gas, collecting such a tax might be very difficult at any point downstream from the point of production--the wellhead or mine mouth--but for oil products it could easily be collected at the refinery or point of importation. In the case of electricity, there is no established price for inputs of coal and hydropower to electricity generation on which an ad valorem tax could be based. If electricity sales were taxed ad valorem, it would be necessary to exempt or rebate taxes on fuels purchased by electric utilities. Without specifying how taxes would be applied to electricity sales and the purchase of fuel by electric utilities, it is impossible to estimate what tax rate (as a percentage of retail price) would be needed to collect any specified amount of revenue.